SUCCESS WITH
Routing

SUCCESS WITH
Routing

STUART LAWSON

GUILD OF MASTER CRAFTSMAN PUBLICATIONS LTD

First published 2005 by
Guild of Master Craftsman Publications Ltd,
166 High Street, Lewes,
East Sussex BN7 1XU

Text © Stuart Lawson, 2005
Copyright in the Work © Guild of Master Craftsman
Publications Ltd, 2005

ISBN 1 86108 490 0
A catalogue record of this book is available from the
British Library.

Production Manager: Hilary MacCallum
Managing Editor: Gerrie Purcell
Project Editor: James Evans

Photography: Anthony Bailey
Additional Photography: Axminster Power Tool Centre,
Microfence, Trend, WoodRat, Ralph Laughton,
John Bullar and Stuart Lawson
Illustrations: Simon Rodway
Cover design: Oliver Prentice (GMC studio)
Book design: Ian Hunt
Typefaces: Palatino and Frutiger

Colour origination: CTT Reproduction, London
Printed and bound: Kyodo Printing, Singapore

Acknowledgements

I would like to thank the following people for their help and support in putting this book together: Anthony Bailey, Ron Fox, Bob Wearing, Ralph Laughton, John Bullar, Simon Rodway, Malcolm Stamper and James Evans.

For Ruth, Oscar and Beatrice.

Contents

Part 2:

Techniques

Part 3:

Projects

Introduction

The router is certainly no ordinary power tool – with jigs and accessories this versatile machine can transform the way you work. So, if you are interested in bench woodworking, are thinking of buying a router, or simply want to get the most from your existing machine, then read on. I can guarantee that by using your router to its full potential, the quality of your woodwork and your enjoyment of the making process will improve greatly.

Although reading this book from cover to cover will help you understand what a router is capable of, the best way to improve your woodworking skills is by mastering the basic operations shown and then getting stuck in by making some of the all-important jigs described at the end of the book (see page 149).

In attempting to build these 'simple' but accurate constructions, you will undoubtedly make some mistakes. However, I can guarantee that by the time you have built a router table and a handful of jigs, you will have an excellent grasp of routing and its potential. You will begin to see the router's handiwork everywhere, and this in turn will inform your own ideas about what to make.

The skills covered in this book will help you design and build what you actually need, with a level of accuracy normally only found in professional workshops. Since MDF and double-sided tape are cheap, there is no excuse for settling for a sub-standard jig or template – strive for perfection and you will be rewarded, both in terms of the knowledge gained and the results achieved.

Crown moulding formed on a large router table using a single crown cutter

Boards tongue and grooved on a router table using a dedicated tongue-and-groove cutter set

Rebate formed using a straight cutter and a router with a parallel fence

Stopped housing formed using a simple square straightedge jig and a straight cutter

Stile/side panel. 'Indent form' routed using a bearing-guided cutter and template

Roundover cutter

Top boards joined using a routed loose-ply tongue slot

Drawer assembly. 'Through' and 'half-blind' dovetails cut using mid-range jig. Slot for bottom routed using a 6mm cutter

Frame and panel door construction. Profile and scribe cutters used to create frame joints and panel slot. Panel-raising cutter used to create panel

Rail biscuit jointed to end panel. Biscuit slots routed

'Foot form' routed using a bearing-guided cutter and template

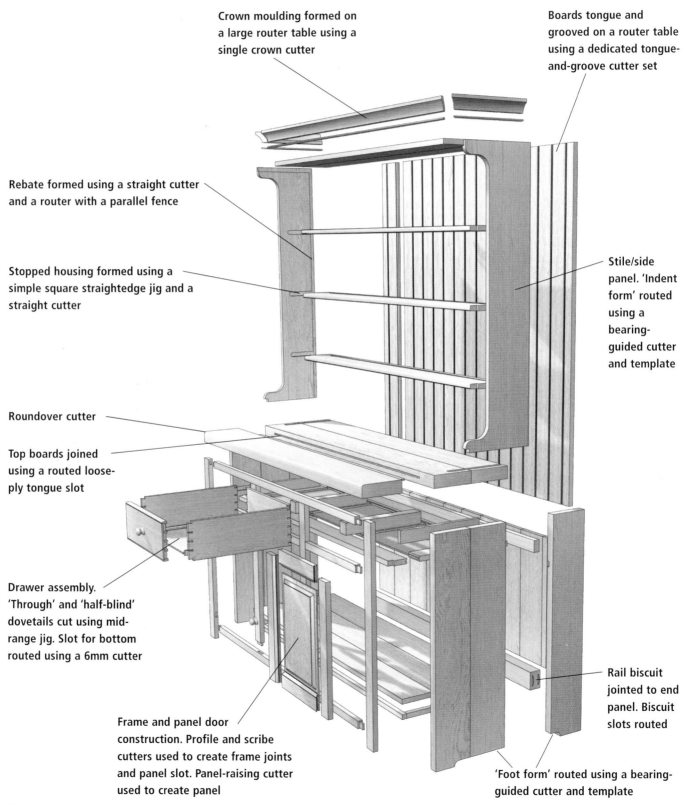

LEFT A small storage chest, also made using a router. Original designed and made by Ron Fox.

ABOVE Apart from the initial preparation of the timber, this dresser was almost entirely crafted with a router. Original designed and made by Ralph Laughton.

Lid trimmed with a bearing-guided cutter (after assembly).

Lid parted from box using a 6mm cutter in a router run against a straight edge.

Hinge recess routed using a simple template and guidebush.

Slot routed to receive a 6mm ply bottom.

'Through' dovetail joints routed using the sophisticated Trend DC400 dovetail jig.

Part 1:
Equipment

1:1 Router anatomy

In the UK and mainland Europe, fixed-base routers have not really been sold or used widely since the 1970s, when, around 30 years after its inception, the plunge router increasingly became the tool of choice. These account for 99.9% of the routers sold today in the UK, with one of the few fixed-base models available being the cordless Porter Cable.

In contrast, the fixed-base router has dominated the North American market. It still accounts for around half of the routers sold and, despite the best efforts of the more versatile plunge router, will probably continue to do so.

Hybrids, such as the DeWalt DW618PK with its interchangeable plunge and fixed-base system, will undoubtedly lead the way and may (if they became available in Europe) have a lasting impact on routing culture.

Base and base plate

In fixed-base routers, a removable motor is clamped directly into the base. The clamping system can be loosened so that the cutting height can be adjusted. Once the router is running, the motor – and therefore the cutter – will remain at a fixed distance from the workpiece.

Plunge-base routers differ in that the motor is mounted on one or more spring-loaded posts, allowing it to be moved up and down in a controlled manner. This means that the cutter of a plunge router can be lowered directly on to a workpiece while the motor is running – something that is not easily done when using a fixed-base router.

Whichever type you use, an open base design aids visibility to the cutter. Unfortunately, by the time you have fitted a dust spout there is very little to choose between most models, even those with integral extraction (see page 25).

Base plates

Most routers feature a Tufnol or moulded plastic base plate, on which the router sits. This matches the shape of the base, onto which it is screwed. The main purpose of the base plate is to act as a barrier between the work and the base, allowing it to slide smoothly and without causing damage to the workpiece.

It is not unheard of for moulded plastic base plates to be warped. This produces instability and can also put the cutter out of square. If you encounter this fault, or even warp in the base casting itself, on a new router, return it immediately to the retailer.

It is common practice to make custom base plates to suit different tasks. However, you should always keep hold of the original factory-supplied version, as this will help you to position the mounting-screw holes on jigs or new base plates or router-table insert plates.

Most if not all routers come supplied with a parallel fence, guidebush, collet, spanner and dust spout. It may well be that your model or make of router will not have a full range of guidebush sizes available. If this is the case, you will need to look at buying and fitting a universal sub-base, which can receive a range of sizes. Sub-bases and guidebushes will be discussed in detail in chapter 1:4 (see page 55).

Collets

The pivotal component of a router is the collet: the small, split metal collar that grips the shank of the cutter and links it to the spindle and motor. Unlike a drill

FOCUS ON:

Fixed-base vs.
Plunge Routers

FOCUS ON:

Fixed-base vs. Plunge Routers

When it comes to buying a router, one of the most fundamental decisions is choosing between fixed base and plunge base. There are no hard-and-fast rules here, and both have advantages as well as disadvantages.

The plunge-base system is certainly the more versatile machine, most notably for its ability to begin and/or end a cut in the middle of a workpiece. Such routers also allow the user to make progressively deeper cuts with great ease simply by changing the plunge depth.

However, plunge routers are more complex and have a greater number of parts than fixed-base routers, which means that they tend to be more expensive. Also, fixed-base routers tend to be smaller and less top-heavy than their plunge-base equivalents. This can make them more stable, as well as easier to use for some operations.

ABOVE A plunge-base router. This machine is a Metabo OFE 738, which is a high-quality lightweight router.

ABOVE The Porter Cable is a cordless fixed-base router. This is a high-quality fixed-speed machine that, like most fixed-base routers, is mainly found in the US.

ABOVE A range of collets and collet nuts.

chuck, which can be adjusted to a continuous range of sizes, a collet will only accept shanks of a fixed diameter. The reason for this is the extreme demands placed upon the collet during use. Not only is routing a high-speed operation, stresses are exerted on the cutter from all directions when machining. It is therefore vital that the collet holds the cutter shank firmly and allows it to rotate without any trace of wobble or vibration.

Bit run-out (i.e. a bit that is not spinning on its central axis) may indicate a bent cutter shank. Alternatively, a rising cutter (i.e. one that starts to come out of a collet when placed under load) may be a sign of collet wear. Both of these faults should be checked for on a regular basis. For safety reasons, it is also essential that the collet and all cutter shanks are kept clean and free from rust.

Basic anatomy

Most routers have largely the same key components, although there can be slight variations in their precise design. The following diagram shows some basic features common to variable-speed plunge-base routers. Certain components, such as the plunge bar and depth-setting turret, would obviously not be present on fixed-base models.

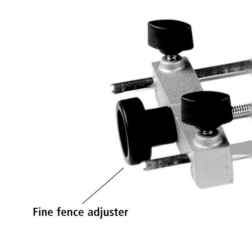

Fine fence adjuster

Speed control

Plunge bar

Twist-style
plunge-lock
handle

Handle

Plunge-bar lock

Collet nut

Trigger/switch

Depth-setting turret

Base

Spindle lock

Parallel Fence

Adjustable fence facings

Because of the range of designs and sizes available, collets are not interchangeable (apart from in some Elu clones). You will therefore need to contact your manufacturer when yours becomes damaged or worn. Since many low-cost router manufacturers do not supply 'spare' parts, you may need to look elsewhere. A comprehensive range of replacement collets is available through major suppliers like Trend in the UK and Eagle America in the US.

The motor

Routers are normally described as ¼in or ½in machines. Although this method of categorizing is now a little old-fashioned, it still applies as general terms. Basically, ¼in routers are the 'smaller' machines, while the ½in variety describes everything from about 1200W upwards (bearing in mind that the output wattage of routers ranges from 700–2300W).

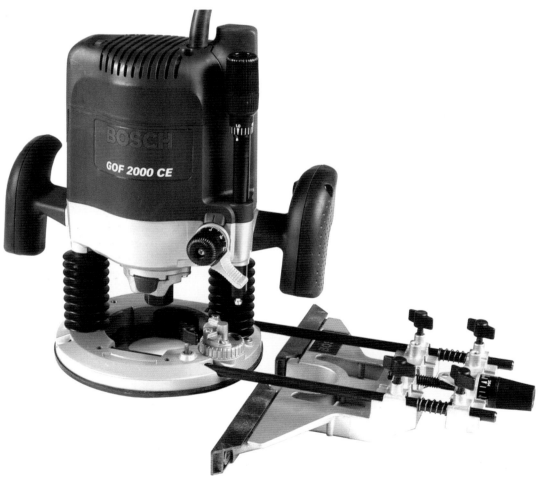

ABOVE The Bosch 2000CE is a large professional-specification router with a powerful motor, and is ideal for inverting in a table (see page 111).

FOCUS ON:

On/off Switches

Health and safety guidelines for amateur machines have largely brought an end to simple locking on/off switches. There is now a variety of designs, all of which require the switch to be permanently depressed whilst routing, which can be particularly problematic when routing by hand or inverting. Supplied straps, cable ties and insulating tape are all used as solutions to this problem.

However, this form of ad-hoc trigger locking should only be attempted when used in conjunction with an NVR switch (see page 67), and must never be used for freehand (upright) routing.

Some professional machines do still offer locking on/off switches, but for newcomers to routing – who will have had little or no training – the safer option provided by amateur machines is entirely appropriate.

ABOVE On/off switches come in a variety of designs, most of which require the switch to be permanently depressed whilst routing. Few amateur machines offer a locking ON position.

FOCUS ON:

On/off Switches

The ideal power for a 'first and only' router is somewhere between 700 and 1300W, but the majority will be between 850 and 1200W. Obviously, the more power and collet sizes you have, the more versatile the machine becomes. However, the increase in size and weight will also make it more cumbersome. A 1100–1300W ½in machine will have enough power for most amateurs' table-mounted jobs, but will be only just large enough for routing lock mortises and so on.

Depth control

Depth control on fixed-base routers is facilitated by a clamping mechanism that allows the motor to be set at a specified vertical position in relation to the base. To achieve the desired depth of cut, the motor is simply unclamped and moved up or down as required.

The depth of a plunge router is adjusted using a plunge bar, which is set in relation to the depth turret. Some models have integral or optional fine height adjusters that make it easier to adjust the router plunge by very small amounts. These adjusters can also be made for

ABOVE Fine height adjusters like this make it easy to adjust the router's cut depth by very small amounts.

ABOVE A six-stage depth-setting turret on a plunge-base router.

other routers whose manufacturers don't offer them as an accessory. These are either connected to a threaded rod in the depth turret or made as a box spanner-type addition to the plunge-limiter rod (which is present on some larger models). Stepped depth turrets are not generally useful unless they include a threaded rod, since this makes it possible to fit a fine height adjuster.

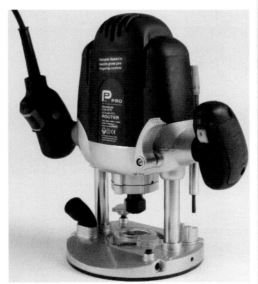

ABOVE Separate locking levers have largely superseded the Elu-style of plunge lock. (On this model, the locking lever is the large silver switch located next to the right handle.)

FOCUS ON:

Handles

Some Elu-style plunge routers have a plunge lock incorporated in one of the handles. However, separate locking levers on the rear of the router body have now largely superseded this twisting-action style of plunge lock. The effectiveness of this design feature is dependent on how easy it is to reach and operate whilst both hands are on the handles.

Indeed, 'ergonomic' grips and integral on/off switches are now pretty much standard, although there was little wrong with the old design! Some newer models also have the facility for adjusting the angle of the handles, but in general this is a useful gimmick rather than a reason to choose one machine make over another.

LEFT Circular Elu-style router handles like these are still featured on several routers.

FOCUS ON:

Handles

Plunge depth

Plunge routers have plunge capacities of between 25–75mm. For most applications, 50mm is the minimum requirement, particularly if you are inverting the router in a table. The advantage of having a greater plunge depth is simple – it enables you to present more of the cutter under the router. This is most important when using a router table (see page 111) or a

(see page 111)

KEY POINT

KEY POINT

It is worth noting that add-on dust spouts/cowls often interfere with the maximum plunge depth of a router, reducing it by anything up to 20mm. This is rarely mentioned in a router's specifications table.

ABOVE A good depth of plunge is essential.

guidebush (see page 97), since the former has to contest with the thickness of the insert plate and the latter the guidebush collar depth.

Very few routers have integral dust extraction, so most models resort to add-on dust spouts or cowls. These are perfectly serviceable, but can restrict the specified plunge depth and occasionally (such as on the Ryobi ERT-1500V) obstruct the fitting of a guidebush. I prefer the vertical or diagonal types of exhaust since it makes it easier to get the extraction hose away from the work area. However, horizontal ports are certainly viable. Dust extraction is covered in greater detail in chapter 1:5 (see page 71).

ABOVE A horizontally venting dust port.

ABOVE A diagonally venting dust port like this makes it easier to keep the extraction hose away from the work area.

1:2 Which router?

The current router market is so competitive that, almost without exception, you 'get what you pay for'. The influx of low-cost routers over the past few years has polarized the market and, whilst making routing more accessible, this has also made the gulf in quality between pro and amateur machines wider than ever. However, competition at all levels of the market has been intensified by the introduction of reasonable quality, budget machines such as the Performance Power Pro range.

Just because a machine looks like another, this does not mean that it will perform and last like those it emulates. The number of cheap Elu 'lookalikes' have been a case in point, since these tend to be 'off the shelf' products that are branded simply by cosmetic changes and sticky labels. Their sellers will rarely know anything more about the router than its product code.

Quality control is normally the first casualty of cost cutting. Following hot on its heels is after-sale service, which for bargain-basement machines is virtually non-existent. If you want a repair under warranty, DIY stores and high-street chains are normally only able to supply a whole new tool. This isn't necessarily as good as it sounds – you might well end up encountering the same problem again. Quality, the provision of repairs, and the assurance of technical advice are assets you pay for when you buy a 'decent' product. Occasionally, it may seem like you are paying over the odds, but if you can afford it and are serious about your woodwork, it is worthwhile spending the extra required for these services.

Choosing the right router

There is a bewildering range of machines available to the first-time router buyer, and it can be tricky if you are relying on the marketing blurb when making your decision. The golden rule is to think in terms of your proposed workload, which will determine the quality you need. You should aim to buy a router of a size and at a price that corresponds with this. The following scenarios should provide some additional pointers.

Scenario 1

Woodworkers who use their machine just a few times a year – perhaps for making the odd picture frame, routing hinge recesses or an occasional worktop – could be well served by a budget ¼in or ½in model. However, if you can afford more and take pleasure from good engineering, there is no reason why you shouldn't consider scenario 2.

Scenario 2

If you are likely to spend most weekends in the workshop and are interested in learning to make furniture, kitchen doors and so on, I would recommend buying a ¼in or ½in standard semi-professional machine. The Trend T5, DeWalt 613K and 620, Freud FT 1000 E and Hitachi M 8V are all good machines. The build quality of this type of router should be sufficient to not hinder your progress, and they should be able to handle a 'heavy' amateur workload. Again, you can spend more if you want,

RIGHT A semi-professional router like the Trend T5 offers good build quality and should be able to handle the workload of most amateurs.

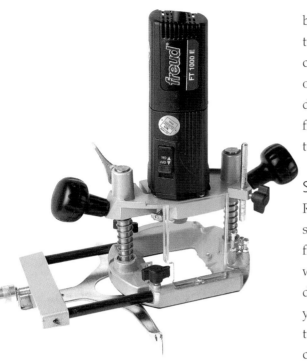

ABOVE A Freud FT 1000 E.

but in my opinion you would do better to use the extra money for buying jigs or cutters. If this category describes your level of ambition but not your budget, you could do well with a 1200–1300W ½in router from the budget range of a DIY store or tool catalogue.

Scenario 3

Rather than being a first choice, I would suggest that this is a natural progression from scenario 2. Once you understand what a router can do and once you have discovered your woodworking 'niche', you should be in a good enough position to be able to make a wise and informed choice about spending enough for a top-quality machine.

ABOVE A Hitachi M 8V.

ABOVE The Makita 3612 is a professional machine that offers a top-quality spec and bags of features. Buying this type of router is all about experience – you need to be able to make an informed decision based on your individual needs and an understanding of what a router can do.

ⓕFOCUS ON:

Amateur vs.
Professional
Routers

ⓕFOCUS ON:

Amateur vs. Professional Routers

The main differences between amateur and professional routers are their comparative usability and the number of working hours they will survive. A joiner's router might be in action for two hours or more every day. In contrast, an amateur set-up would be unlikely to make such demands in an entire week. Obvious stuff really, but it does go a long way to explaining the difference in price between, for example, a Ryobi and a Festool. Neither of these models is over or under-priced – their relative costs simply reflect the time and care taken during the manufacturing process, as well as their different countries of origin.

ABOVE When your proposed workload is relatively light, a budget machine like this Ryobi ERT-1500V represents an adequate, affordable option.

ABOVE The Festool OF 2000 comes at a price, but this large professional router offers lots of power, smooth handling and good build quality. If you require high-calibre results from demanding applications, the extra expense will provide the necessary quality and durability.

Low-cost routers

Limited funds need not get in the way of your woodworking ambitions, but it is important to have realistic expectations of what a low-cost tool can do. The following list of common difficulties and faults encountered with low-cost routers will help you establish whether you can blame your tools for poor results.

(KEY POINT

It is worth noting that the manufacturers of many low-cost routers do not supply any spare parts. You should therefore check that a specialist supplier could supply additional collet sizes or other parts before buying a bargain-basement router of unknown branding.

(KEY POINT

ABOVE Low-cost, off-the-shelf machines such as this Power Pro CLM 2050W offer reasonable quality for anyone on a budget. If you opt for a router of this type, be realistic about what it can achieve and don't expect great after-sales service from the retailer.

Poor guidebush alignment

Most routers of reasonable quality will be manufactured to a high enough standard to enable the guidebush to sit centrally to the cutter (not 100% accurate, but good enough). However, this potential fault can affect the quality of your work because you might experience a slightly different set-up position every time you fit the guidebush. Alternatively, it might be set at a slightly different position from each side of the base.

I would always recommend using a guidebush-centring device (see page 66), whatever the quality of your machine. Of course, if you have a more expensive Bosch or a Festool router, you will have the advantage of a quick-fit locating system, but I would still check that these are accurate when you buy the router.

Solution: If there is not enough 'play' in the guidebush fixing position to adjust it with a centring device, it may become necessary to drill some slightly oversized holes in the guidebush rim.

Column wobble

One of the most common faults in all fixed-base routers is a degree of movement or wobble between the motor body and the plunge columns. Disappointingly, nearly all routers have some movement, but if your machine moves the end of a long cutter even 0.5mm out of alignment you will have problems with accuracy. Cuts will often not be exactly square or repeatably accurate.

Solution: This is a major problem that will significantly affect your work, so either get a replacement machine or a full refund.

Plunge-turret movement

Whatever the design of plunge turret, there is often some 'play' in the block when depressed from one side. Problems arise when this movement makes it possible for repeat cuts to vary by significant amounts (sometimes up to 2mm). Even though I was aware of the problem whilst testing low-cost machines, I was still caught out by this design fault.

Solution: This is an annoying build fault that requires further investigation. It may of course be an 'as standard' feature, in which case you can either account for the problem every time you use it or you can try to get your money back and buy a different router.

Insufficient plunge 'grip' when inverted

If you are relying on the grip of the plunge lock to secure the height of an inverted router, you may encounter the problem of the cutter being gradually or dramatically drawn away from the workpiece. If noticed, this can easily be rectified. However, if the changes are subtle and not discovered until a job is well underway, the consequences can be disastrous.

Solution: Removing the plunge spring will make raising a lot easier and may improve the grip on the lock. However, the most sure-fire way of overcoming this problem is to buy or make a raising device (see page 63).

ABOVE The DeWalt DW626 plunge router has a powerful 2300W motor and a ½in collet. This professional-level router will tackle the most demanding applications, and can be used in either freehand or inverted (i.e. table) operations.

Ⓕ**OCUS ON:**

What Size?

Ⓕ**OCUS ON:**

What Size?

As explained on page 20, routers are normally described as ¼in or ½in machines; ¼in routers are the 'smaller' machines and ½in describes everything from about 1200W upwards. As you can see from the table below, larger routers are mostly able to receive the smaller sizes of collets (and therefore cutter shanks) as well.

The majority of router users start out looking for a 'jack of all trades' machine, which by its very nature will be a reasonably priced, simple yet complex, lightweight, powerful, top-quality piece of German engineering!

In order to make this perplexing task of choosing the right router a little easier, the following table should help you make up your mind.

Most professional workshops will have two or three routers under the bench. If you have the funds and want to emulate this 'ideal' set-up, there are various possibilities. If you are aiming for a two-router option, try an 800–1100W and a 1600–1800W machine. For a three-router set-up, go for ratings of 700W–1000W, 1100–1600W and 1700–2300W.

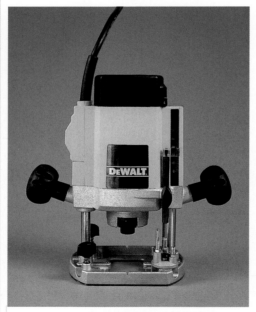

ABOVE The DeWalt DW 615 is a lightweight plunge-base router with a 900W motor and ¼in collet.

Power	Example router	¼in Collet	½in Collet
500W	Trend T3 (550W)	•	
700W	Metabo OFE 738 (710W)	•	
900W	DeWalt DW 615	•	
1100W	Makita PR1110C	•	
1300W	Power Pro CLM 1250RD	•	•
1600W	DeWalt DW 624	•	•
1700W	Trend T9 (1800W)	•	•
1900W	CMT 1850W	•	•
2100W	Bosch 2000CE (2000W)	•	•
2300W	DeWalt DW 626	•	•

Scope for fine height adjuster fitting

The lack of an integral threaded rod in the depth turret can mean that the traditional solution to fine height adjustment is made impossible. However, if you have a plunge limiter, you can then adapt a box spanner to attach to the end of the limiter rod and act as a more coarse, but still very useful, fine height adjuster. Alternative solutions such as a RouterRaizer (if appropriate) or a home-made device fitted under the router body may also become necessary.

Solution: Seek out a raising device that has the potential for fine adjustment. A car jack is a popular low-tech solution and would solve both problems (although the plunge lock must be released first).

ABOVE The CMT 1850W with its associated kit.

1:3 Cutters

Despite the vast amount of extra kit that is available for routing, the ultimate routing accessory is a cutter. You can be sure that no matter how many cutters you own, you'll always be looking for an elusive profile and shank-size combination. Accordingly, it is always better to design projects based on what you have and be resourceful.

It is important to buy a cutter of an appropriate quality for your expected workload. Buying an expensive cutter may not be appropriate if it is for a rarely used application. Equally, poorly manufactured cutters could blunt easily, be inaccurately made or possibly even dangerous. Unbalanced cutters, poor brazing and low-quality tungsten carbide can all add to the problem.

My policy is only to use cutters whose manufacturers achieve a quality standard for at least some of their cutter ranges. Trend's Professional range of cutters, for example, passes the European Standard EN847-1 (which supersedes the Holz BG Test) for quality of manufacturing and materials. While their Trade and Craftsman ranges do not achieve this standard, my view is that a responsible manufacturer will be diligent from the top to the bottom – they will not risk their brand reputation with inferior and possibly dangerous cutters, even in their budget ranges.

⊤ECHNIQUE:

Fitting Cutters

⊤ECHNIQUE:

Fitting Cutters

Cutters are fitted into a router by inserting their shank into a collet, and this in turn is located in the router shaft. Both are held in place by a collet nut, which tightens the collet onto the cutter's shaft. The diameter of a shank relates exactly to the size of the collet, and collets are designed for specific shank sizes and routers. It is therefore vital that you have the correct collet size for your cutter shank – 12mm may be close enough to ½in on a building site, but not in routing. Some collets are able to receive two or more collet sizes, which can be a useful feature. Finally, when fitting a cutter into the router, always make sure that at least half the shank (to a minimum length of ¾in or 19mm) is inside the collet. This will give adequate support and prevent stress to the shank.

LEFT Fitting a cutter. The shank should be inserted at least halfway into the collet to ensure that the cutter is centred and held securely.

LEFT Simply engage the spindle lock and tighten the collet nut to hold the cutter in place.

The cutting edge

There are five types of material used in the production of router cutters, although only two of these should concern the average routing enthusiast. These are tungsten-carbide tipped cutters (TCT) and, to a lesser extent, high-speed steel cutters (HSS). The remaining types are: super high-speed steel, solid tungsten carbide and poly-crystalline diamond (which is manufactured by bonding tiny diamonds to the surface of tungsten carbide).

There are some advantages to using the latter types of cutter, such as high levels of performance and excellent durability. However, all three are really designed for professional and industrial applications. For example, super high-speed steel cutters, which are manufactured using a specially developed high-grade steel, are typically used in industrial applications such as machining aluminium.

KEY POINT

A well-designed cutter will be efficient at clearing waste and reducing friction by giving clearance between the body of the cutter and the workpiece. However, a cutter's performance is only as good as its user – feed rate and cutting depth (see page 94) are of paramount importance.

FOCUS ON:

Shank Size

The most common sizes of shank in the UK and US are ¼in (6.35mm), ⅜in (9.5mm) and ½in (12.7mm), whereas in mainland Europe 6mm, 8mm and 12mm are the norm. Because the imperial sizes are so much more prevalent, metric-shanked cutters are less common and have a smaller range of profiles available. Cutters with ⅜in and 8mm shanks are not so widely available, but they do offer the owners of smaller routers the opportunity to use larger cutters.

The size of the shank has a major impact on the potential size of the cutting edge (i.e. the larger a shank diameter, the larger its cutter 'head' can be). This is simply because of the stresses exerted on the shank whilst machining. It is not difficult to over-stress a cutter and shear it off just below the head. A bigger shank is stronger, and will therefore better resist bending or breaking when in use.

High Speed Steel (HSS)

Some low-cost cutters are made from HSS, with the best quality bits of this type being machined from a single piece of steel. There are several advantages to using an HSS cutter: they are cheap, are particularly well suited to machining softwoods, and can be hand sharpened using a small diamond stone.

FOCUS ON:

Shank Size

KEY POINT

ABOVE An example of a high-speed steel (HSS) cutter. This type of cutter is cheap and can be used for machining softwoods.

ABOVE A tungsten-carbide tipped (TCT) cutter. These hardwearing bits are the most widely used type of cutter.

However, HSS has been largely superseded by the more versatile, and more durable, tungsten-carbide cutting edge. This is because, although an HSS cutter can be honed to a razor sharp edge, it is quickly dulled on hardwoods and man-made boards such as MDF.

Tungsten-carbide Tipped (TCT)

Tungsten-carbide tips (otherwise known as 'blades') are brazed to a high-grade steel cutter body to produce what is the most widely used type of cutter. Although tungsten carbide is relatively brittle, it is also extremely hard. As a result, despite not being able to achieve as sharp an edge

as HSS (because TCT is made from sintered tungsten-carbide granules), TCT cutters stay sharp for longer. Man-made boards and, in particular, hardwoods present no problem to this universal cutter.

(KEY POINT

(KEY POINT

TCT cutters can be honed using a small diamond stone. Because the stone will remove a small amount of tungsten carbide, any damage to the edge will need to be addressed by a professional tool grinder.

Cutter types

There are several different main cutter types, which relate to distinct types of cuts and applications. Amateur router users will mainly have cause to use arbor-mounted, bearing-guided, pin-guided, single and two-fluted cutters, but as with most things technical, it always pays to be aware of all of the options available.

Arbor-mounted cutters

One or more slotting cutters mounted horizontally on an arbor. These combinations are either used for slotting frames for panels, tongue and groove joints, biscuit jointing and for profile and scribe joints. These types of cutter can often be taken apart and reassembled using different bearings, washers and spacers in order to produce different heights and depths of cut.

ABOVE Arbour-mounted cutter.

Bearing-guided cutters

These 'self-guiding' cutters are able to follow the edge of a workpiece or template by the use of a rotating bearing guide. For trimmers this is of the same diameter as the smallest part of the cutter, whereas for rebating sets the offset between the bearing and the cutter diameter equals the width of the rebate.

ABOVE Bearing-guided cutter.

Multi-profile cutters

A replaceable-tip system that uses a range of single-blade cutting profiles within two head types, either bearing-guided or panel.

Pin-guided cutters

The pin at the base of the cutter works in the same way as a bearing guide but rotates with the cutter and fits into smaller spaces. In use, care is needed since the friction of the pin has a tendency to cause

ABOVE Multi-profile cutter.

ABOVE Pin-guided cutter.

burning and 'biting in'. Pin-guided cutters are less common than bearing-guided cutters for the simple reason they produce less predictable results.

Replaceable-tip cutters

Mainly used in CNC (Computerized Numerically Controlled) routing applications, replaceable-tip cutters utilize removable tungsten-carbide blades secured with a Torx key to a high-grade steel body. The blades can often be rotated up to four times (for a straight-sided square cutter). They reduce down time in CNC applications, which has an impact on production costs. For regular routing applications, they can be cost-effective for high-volume work.

ABOVE Replaceable-tip cutters.

Single-fluted cutters

Single-fluted cutters are stronger by design since only one flute has to be machined from its body during production. Typically they are small-diameter straight and engraving cutters whose design enables a faster feed rate, although this in turn reduces the relative quality of cut.

Spiral cutters

Designed with similar helical flutes to a drill bit, these fast-cutting up-cut or down-cut router cutters are mainly used

ABOVE Single-flute cutter.

FOCUS ON:

Cutter Sets

FOCUS ON:

Cutter Sets

The cutter selections offered by manufacturers are a good way to get started, although there will probably be one or two cutters that you never use in each set. If you are confident about what you need, you can probably save money by buying cutters separately. However, it is still worth seeing what cutter sets have to offer in order to give you some ideas.

Of course, the low cost of some unknown-brand sets will be tempting for those with limited resources. I would advise against buying them if at all possible. Whatever your choice, eye protection and tough work clothes are a must.

If you are fortunate enough to have a ¼in and a ½in router, the contents of the two quality starter sets shown would be ideal. Alternatively, here are my recommendations for a basic cutter selection:

¼in shank

1. ¼in (or 6mm) straight, two-flute cutter
2. ⅜in (or 10mm) straight, two-flute cutter
3. ½in (or 12mm) straight, two-flute cutter
4. ¾in (or 18mm) straight, two-flute cutter
5. ½in (or 12mm) bearing-guided trimmer. Long (bottom bearing)
6. ½in (or 12mm) bearing-guided trimmer. Low profile (bottom bearing)
7. ½in (or 12mm) radius cutter
8. Bearing-guided bevel cutter
9. Bearing-guided ogee
10. Bearing-guided ovolo
11. Dovetail cutter (for dovetail housings)

½in shank

1. ½in (or 12mm) long straight cutter
2. ¾in (or 18mm) short straight cutter
3. ½in (or 12mm) long bearing-guided trimmer (bottom bearing)
4. Bearing-guided bevel cutter
5. Bearing-guided ogee
6. Bearing-guided ovolo

LEFT Boxed sets of high-quality, semi-professional ½in and ¼in cutters by Titman.

ABOVE Spiral cutter.

ABOVE Two-flute cutter.

in CNC applications. Although they can also be used more widely to prevent surface breakout, their high cost could discourage many amateur users from putting this to the test.

Two-fluted cutters

These are the most commonly found cutter types, featuring two cutting edges on opposite sides of the cutter. For plunge cutting, an additional cutting edge is ground on the blades and in some instances an additional third blade is brazed on to the bottom. This can be positioned either in line or at 45° to the other two blades.

Cutter Profiles

There is a huge range of cutter profiles available. The only restrictions on their use are the size of your router, its available collets and whether you can invert it in a table. A great many routing operations and cutters are best applied in a table set-up, particularly with cutters of above 50mm diameter. Large panel raisers, for instance, should never be used in a hand-held router. This section will give you an understanding of the types of profile available, but for a comprehensive reference guide you'll need to order a selection of cutter catalogues.

ABOVE A selection of cutters showing the huge range of profiles available.

Beads and reeds

A bead is a concave or convex 180° decorative roundover profile, most commonly featuring a shoulder. A reed is simply a multiple of these 'stacked' on one cutter.

ABOVE Reed (left) and beed cutters.

Bevel, Chamfer and V-groove

Strictly speaking, a bevel is an angle along the entire side of a workpiece, whereas a chamfer is formed by the removal of a 45° angle between two faces. Typically, a bevel is a construction detail and a chamfer is a decorative feature. Bevel cutters are available in a range of angles (usually 45°, 60°, 67.5°, 70°, 75° and 80°) and are mostly used to create multi-sided constructions such as a coopered chest lid.

Cornice

Strictly for use in large, inverted routers, cornice moulders have profiles of up to around 60mm. Cornices do not have to be machined with a single cutter; they can just as easily be built up as composites using several cutters. Whichever way you approach its production, however,

ABOVE Two bearing-guided bevel cutters.

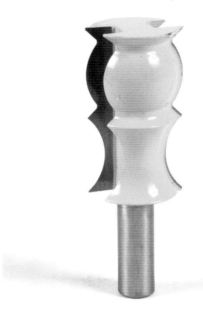

ABOVE Cornice cutter.

the ability of a cornice to transform a tall cabinet or dresser into something special is beyond dispute.

Dovetail

The number of dovetail jigs on the market has been increasing year on year, as has the range of cutters. In addition to being used for cutting traditional end-to-end dovetail joints, this type of router bit can also be applied to sliding dovetails and used to cut stair housings, although for the latter their pitch is not as steep. Specific jigs require specific cutters, and so despite the fact that one is included in most boxed sets, they will only be of use for dovetail housings.

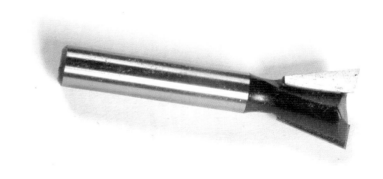

ABOVE Dovetail cutter.

Down shear and up-down shear

These are designed to eradicate break-out and surface spelching respectively. The former is good for grooving and for trimming laminated boards and veneers.

Drill cutters

A router is the ideal solution for certain drilling operations – it is extremely fast and the holes are guaranteed to be vertical. Using a standard HSS drill bit, even one of a related collet size, is not recommended as it may snap off in use. For drilling perfect dowel holes several manufacturers produce lip and spur bits for routers.

ABOVE Jointer.

Jointers

There are four types of cutter within this category: finger jointers, tongue and groovers, corner mitres and mitre-lock jointers (which can also be used for a glued-edge joint). None of these actually produces a self-holding joint – all require additional gluing and clamping. Despite being manufactured accurately, all of these routed joints are only as good as the set-up and preparation of the timber, which should be square and cut to length. Always set up the cuts with test pieces before machining a workpiece.

Laminate trimmers

Whilst a standard bearing-guided trimmer can make a good job of trimming the edge of a bonded laminate, there is a whole range of cutters dedicated to variations of this job. Versions include unguided flush trimmers, bevel trimmers, pierce and trim cutters and those that will cut two parallel laminates at once.

ABOVE Up-down and down-shear cutters.   **ABOVE** Router drill cutter.

ABOVE Laminate trimmer.

Linen-fold sets

This is a relatively new development from Wealden Tool Company that enables router users to recreate the classic, previously carved, linen-fold panel decoraton. One cutter is used to machine the linen profile, after which, with the use of a template, a smaller bit cuts the 'openings' of the folds.

ABOVE Bearing-guided ogee cutter.

Ogee

A classic decorative S-Shaped moulding profile incorporated into panel moulders and bearing-guided bits.

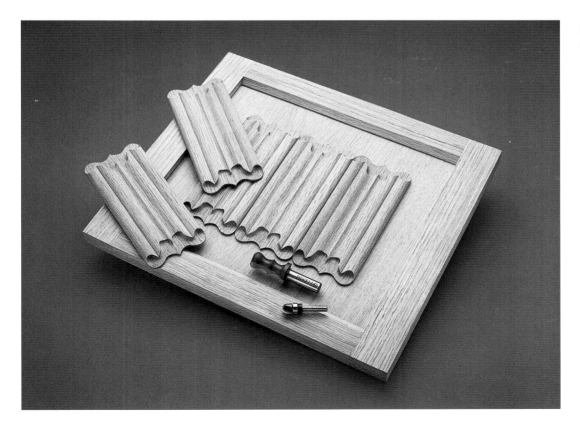

LEFT The linen-fold cutter set-up.

ABOVE Bearing-guided round-over cutter.

ABOVE A very large panel raiser.

Ovolo and Round-over

This is another classic decorative moulding profile, which can be incorporated into panel moulders and bearing-guided bits. Very simply, it rounds an edge to a given radius. It is also possible (if used with a good set-up) to round over two edges to produce a radiused edge.

Panel Raisers and Moulders

Panel raisers are used to profile a decorative and functional tongue around the edges of panels on frame and panel doors. Horizontal panel raisers are some of the biggest cutters around, and are almost exclusively ½in-shanked cutters. Medium-sized versions can be used in an upright router, but all are best employed in a table set-up. The vertical-cutting varieties are available in smaller shank sizes, making them more accessible to hobbyists. Panel moulders have a continuous profile around the sides and base and are either used for faux raised panelling or for adding to multiple-profile workpieces such as wide architraves.

Pocket

This type of cutter is used for deep grooving and mortising applications. An extra-long shank means that the cutter suffers proportionally more stress in use, and so a greater number of passes, at less depth than usual, are required.

Profile and Scribe

There are two types of profile-and-scribe cutter: combination and all-in-one cutters. Combination cutters must be disassembled in order to cut both parts of the joint. All-in-one cutters, on the other hand, are taller and this extra height enables them to cut both parts of the joint simply by being raised or lowered.

The traditional construction of frame and panel doors has been made easier by the use of profile-and-scribe cutters. By using and reassembling the same cutter to profile and machine the joints of the door, perfect results are possible every time – with the right set-up. Panel-door sets include a panel-raising cutter and a profile-and-scribe bit.

Radius, Cove and Cavetto

Classic decorative moulding profiles incorporated into two-flute, radius, panel moulders and bearing-guided cutters.

ABOVE Pocket cutter.

ABOVE Radius cutter.

ABOVE Profile-and-scribe cutters.

ABOVE Bearing-guided cove cutter.

ABOVE Rebating set.

ABOVE Bearing-guided slotter.

Rebaters

Rebating cutters can be bought with one or more bearings. The size of the bearing in comparison to the diameter of the cutter determines the width of the rebate. Of course, the width of a rebate can be determined by the use of a fence or parallel guide, but for applications such as rebating assembled frames, the rebate follower comes into its own. Rebating sets can have anything up to seven different-sized bearings with one cutter, enabling the user to make either machine-stepped rebates or a range of widths around a bearing-followed edge.

Slotting and Grooving

These are arbor-mounted saw-like cutters used for a wide range of applications including tongue-and-groove joints, biscuit-joint slots, panel slots, plus intumescent and weather seal-receiving grooves. Some arbors can take multiple

cutters; used in conjunction with spacers and shims, these give very particular and useful results.

Straight

'Straight' is more of a general heading than a reference to one particular cutter, as it includes single-fluted and two-fluted cutters. Single-fluted cutters are stronger because only one flute has to be machined from the cutter's body during production. They are typically small-diameter straight and engraving cutters whose design enables a faster feed rate, although this in turn reduces the relative quality of cut.

Two-flute cutters are the most commonly found cutters. They feature two cutting edges on opposite sides of the cutter, with an additional cutting edge ground onto the bottom of the blades to enable plunging.

ABOVE Two-fluted straight cutter.

ABOVE Stagger-tooth cutter.

In some instances, an additional third blade is brazed to the bottom either in line or at 45° to the other two blades.

Stagger tooth/mortise

The two staggered blades provide improved chip clearance in deep mortising applications, but also produce increased vibration. In the same way as a pocket cutter, the extra-long shank means that the cutter suffers proportionally more stress in use. Therefore, a greater number of passes, at less depth than usual, are required.

Because this type of cutter is not suitable for plunging, a ramping technique should be used. I would advise novices to use straight cutters instead, and thus avoid this additional complication.

Trimmers

Depending on the size of the bearing, this cutter can be used for trimming one surface either flush or parallel with another. The majority of these 'template followers' have bottom bearings, but some feature top or even top and bottom bearings. V-groove trimmers are used to flush-trim and hide a glue line in one pass.

ABOVE Trimming cutters.

1:4 Accessories

Routing surely has more associated kit than any other area of woodworking. Therefore, in an attempt to make searching a one-stop exercise, this chapter provides a breakdown of the kit that is available and what it does. I have excluded cutters as these are covered elsewhere in greater detail (see page 37).

As you may know, many jigs are easily made in the workshop. However, there is an increasing number of innovative devices whose manufacture requires top-notch engineering. These can significantly improve the lot of the router user, and although some are most definitely luxuries, others have the potential to support and improve the quality of your work without breaking the bank.

All of the products featured here attain a very reasonable quality standard, but this will vary and inevitably some are better than others. The best thing is to find out what works for your needs and your budget.

Clamp guides

Whilst there is nothing wrong with using a straight edge or length of timber, clamp guides are an excellent innovation. They aid cuts with routers, circular saws and even biscuit jointers. They can also be used as a board cramp, although the initial pressure should probably be applied with a sash cramp.

Trend produces excellent work guides that are manufactured to a very high quality, and further attachments such as jaw pads and squaring attachments improve their performance still further. A good basic set-up would be to own the 610mm and the 1270mm guides, although 305mm, 610mm, 915mm and 1270mm guides are also available. Perform Clamp Guides offer a simple, cheaper version that uses a lightweight box-section rather than Trend's quality extrusion.

Routing aids such as clamp guides have partly superseded the use of T-squares (although the latter is still a good piece of kit). However, the ultimate in router guides is the Pro Track system. This involves fixing a router to a base plate, which in turn runs along a straight, flat extrusion to produce perfect cuts every time.

KEY POINT

KEY POINT

When buying a new accessory, always check with the manufacturer or supplier that it is compatible with your router. If there is any uncertainty, reserve your right to return incompatible goods.

ABOVE Simple guides such as these can be clamped across boards to provide a firm, straight edge that a router can be moved along.

RIGHT When used with the correct set-up, the Trend Pro Track can produce perfect cuts every time – facilitating almost foolproof routing.

Collets

Collets are designed for specific shank sizes and routers. The full range of collets available for your router may have been supplied when you bought it. However, if you require some additional sizes or need to replace a collet because it has become worn, you can either contact your manufacturer direct or get in touch with a supplier of routing accessories.

Collet extensions extend a router's shaft and are an essential part of table routing. This is because cutter shanks are often too short and routers can have insufficient plunge depth to present the cutter properly through the table. Both the CMT

and Trend extensions provide increased length (up to 50mm) and conversion (to ½in, ¼in and 8mm) from ½in collets only.

An expensive but well-engineered piece of professional kit is the Xtreme Xtension. This is heavily built and designed to reduce vibration, particularly when using large cutters. The Xtreme Xtension also has dynamic balance adjustment. It is intended for use with variable-speed plunge routers in table applications, and should not be used with fixed-base, single-speed or handheld routers.

KEY POINT

Collet extensions must be manufactured to a high standard and even the smallest of inconsistencies will produce vibrations. If you encounter this problem, do not hesitate to return the product.

KEY POINT

ABOVE CMT collet extensions can be used to provide extra cutting depth, making them especially useful when table routing.

BELOW The Xtreme Xtension is a useful accessory designed for use with inverted variable-speed plunge routers. As well as extending the collet and so increasing cutting depth, it facilitates quick and easy cutter changes.

Jigs

Although a rather general heading, jigs are the accessories – shop-bought or otherwise – that interact directly with a router to perform a function such as circle cutting, preparing lock mortises or simply cutting a hinge recess.

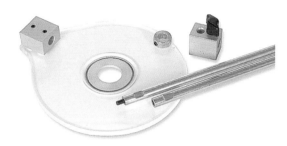

ABOVE A circle-cutting trammel

The circle-cutting trammel is a sub-base and trammel-bar system that will need to be drilled to fit your router. The cutter aperture has room for cutters of up to 60mm diameter, and it is capable of cutting a circle of up to 3m (10ft) in diameter.

An affordably priced large circle and ellipse jig is that produced by Trend, which is able to cut circle diameters or ellipse axis of between 550 and 1800mm (22–71in). It is a well-made product and accurate if set up correctly and used with care. A compact version, which is able to rout a minimum minor axis of 160mm (6in) and maximum major axis of up to 580mm (23in), is also available.

BELOW As well as producing accurate ellipses using a simple connector plate and slide bar, an ellipse jig can also be adapted to cut circles.

A large compass jig is easy to adjust and straightforward to use. It can be made in the workshop, in which case the use of transparent plastic has obvious advantages. A manufactured jig worth considering is the N-Compass. This allows for easy fine adjustments to radii and is useful in conjunction with extra-long rods.

The Accurate Guide, from the US, is a simple piece of kit that gives the user complete, accurate control over housing widths. Whereas previously a housing for some types of 18mm MDF would have been cut with an over-sized imperial version (¾in or 18.2mm) or with a shim added to the fence, this guide enables you to cut and finely adjust your housings by inserting board offcuts to set the width.

The hinge jig is a large adjustable template for the routing of standard butt hinge recesses. Once you have got over the expense, these jigs make short work of what is really a basic woodworking task. A more basic 'Hinge Setting Jig' is available

ABOVE Trend N-Compass

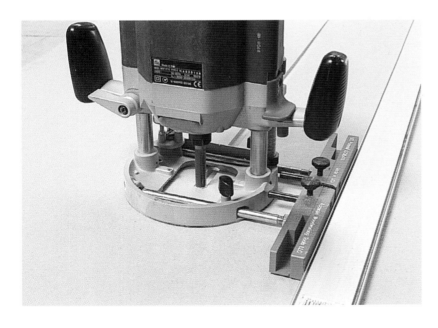

LEFT An Accurate Guide gives perfect results and perfectly tight joints every time, and is also very good for cutting sliding dovetails.

with fixed 3in and 4in apertures, although this does come with a 'free' corner chisel for squaring the recesses.

So many front doors have bodged, jigsaw-cut letter slots that I would always recommend using a router for this job. The well-made Trend jig produces great results easily, but if the price tag makes you baulk, much cheaper budget versions are available from various manufacturers. Alternatively, it is not hard to use a ½in

cutter with a 30mm guidebush and a home-made MDF template.

The Routabout from Trend is an excellent tradesman's tool for creating access hatches in chipboard or ply floors. It comes supplied with a cutter of your choice and spacer rings for three hatches.

Despite becoming easier with practice, cutting out lock mortises is always a time-consuming job with auger bits

ABOVE By using the right jig, the speed and convenience of a router can be brought to bear on those time-consuming, repetitive jobs. For example, a lock-jig kit like this will make cutting out lock mortises a much less demanding task.

and chisels. By the time you have bought the cutters and a guidebush, a lock jig will not be cheap; but if your time is money, this kit will seriously help. Another expensive but useful item is the worktop jig. If you are joining a kitchen worktop the cut has to 100% accurate, and with these jigs anyone can achieve it. If money is no object, it is also possible to get a 1.2m (4ft) long super jig with cutouts for cable tidies and more besides.

A lathe jig used in conjunction with a router makes it easy to produce dowels and decorative spindles. Complex barley twists and fluted columns are simple to produce using this system, for example. Another labour-saving option are stair-housing jigs, which are widely used in the trade. These Tufnol template jigs are virtually foolproof – used in conjunction with a guidebush the jig is simply slid down the string using an integral fence.

ⒻFOCUS ON:

Work Holding

Work holding does not have to be a hi-tech operation – I managed for years with only G-cramps, and there is no reason why you cannot stick to tried-and-tested tools like this. Quick-fix cramps like the Solo and 'F' varieties are often best for work and jig holding.

A more recent development is vacuum work holding. Vac Pots offer versatile, individual work-holding suction cups. These can be used either with Air Press's vacuum pump or in conjunction with a compressor and Venturi valve.

The MiniMach is a highly recommended piece of kit. Powered by a standard dust extractor, this excellent clamping bed is ideally suited to small workpieces. Alternatively, try the UMach and BigMach kits. Following the same principles as the MiniMach, these larger work holders are sold as kits and can make vacuum beds up to 259 × 482mm and 1120 × 520mm respectively. (Note: you will need a dust extractor/vacuum cleaner capable of extracting 6 cubic metres per hour to use these sets.)

ABOVE By using suction from a dust extractor or standard vacuum cleaner to hold down the workpiece, clamping beds provide simple, obstruction-free work holding.

ⒻFOCUS ON:

Work Holding

Safety and dust extraction

Dust extraction from the router and ambient air filters for airborne particles are essential parts of routing kit, so treat them as a necessity rather than an accessory. Apart from reducing the health risk, good extraction also means that you can keep a tidy workshop.

Designed to remove the dust and debris of routing at source, a vacuum extractor is the first line of defence for a router user. The best extractors have automatic switching so that they only turn on when the router is started. Before buying, check that the pipe adapter supplied fits your router dust port. If not, you can always get hold of a suitable (shop-bought or home-made) reducer – or simply use some duct tape.

KEY POINT

Very fine dust particles, particularly those from MDF, are potentially very harmful to your health, and so every measure must be taken to remove these from the atmosphere before they reach your lungs. Ambient filters simply clean the workshop air and provide an important line of defence against the finer dust particles. Make sure you buy a filter of appropriate capacity for your workspace.

Whilst flat facial respirators and cup masks offer enough protection in conjunction with vacuum extractors and/or air filters, they are not suitable as a stand-alone solution. If you are working with no other protection or are particularly sensitive to dust, a self-contained respirator is the only solution.

Safety glasses, goggles or visors are an essential part of routing kit and should be worn at all times when machining. Ear protection is desirable for hobbyists and essential for professionals or those who already have some damage to their hearing. The best combination is a pair of ear defenders and some goggles with an elasticated strap, since neither interferes with the wearing of the other. (For more on safety equipment, see chapter 1:5, page 71.)

KEY POINT

In order to prevent the hose of your dust extractor from snagging and interfering with your work, attach it to the ceiling or a cross beam using a bungee cord or piece of rope.

Parallel-fence systems

The basic parallel fence that is included with every router can become much more useful by using both longer rods and a trammel point. Also, if you can upgrade to a fence with a fine adjuster, you will achieve much greater control over your cuts and not have to wrestle so much over accuracy.

Extension (Mod) rods are excellent accessories for extending the reach of a parallel fence. 300 × 8, 10 or 12mm diameter rods are available, and each are threaded to join to the next so that they can be used in combination with others.

ABOVE For hand-held routing applications in which precision is essential, a MicroFence is the ultimate solution. Compatible with a huge range of routers, this versatile accessory gives you complete control and provides unbeatably accurate measuring.

Extra-long rods are excellent accessories for use with trammel and circle jigs, particularly when used with Beam Trammel Attachment points. Either buy them from a supplier or they are available from an obliging metal merchant (tell them what they are for to make sure you get a decent-quality steel).

The MicroFence system provides the ultimate in accuracy, whether for the minute adjustments available from the MicroFence or for the totally accurate ellipse jig. Few people apart from top-notch cabinetmakers need this kind of kit, but if you do take the leap then you can be sure you have bought the best.

A basic piece of kit but still particularly useful is a trammel. This can be bought either as a ready-made bar with integral pin or as an attachable point for your standard or longer (Mod) 8mm or 10mm rods. Make sure that you check the diameter for compatibility with your router model.

Depth/height control

Dedicated fine height adjusters for plunge routers are widely available from various manufacturers. Essentially, they make it easier to adjust the depth of cut by very small amounts by overriding the machine's plunging action. Fine height adjusters can be used when routing by hand, but the extra control they provide becomes really useful when the router is inverted in a table (see page 111).

Dedicated fine height adjusters are sometimes available from manufacturers, but if your router does not have a threaded rod integral to the height-setting turret, you cannot use them. However, it is often possible to make your own from a box spanner. This is a relatively cheap option and provides a well-engineered solution. It can either connect to the threaded rods in the plunge turret or, if you have one, to the depth limiter.

The PlungeBar depth sets a router smoothly with just one hand – whether hand-held and upright or inverted in a table. This means that you can make a deep cut in several passes very easily. There is a PlungeBar for virtually every make of router and similar products are available from different manufacturers.

The RouterRaizer is a very useful accessory for height adjustment in table-mounted routers. It allows excellent adjustment from above, although fitting it may well involve taking your router apart and drilling a hole in its base.

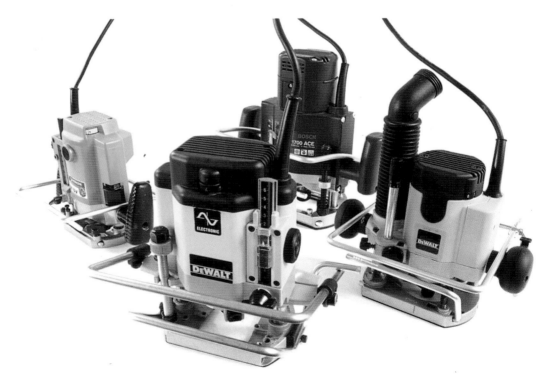

ABOVE The WoodRat PlungeBar is a very easy accessory to use – the router is simply raised and lowered by squeezing the bars together, and this can be done with one hand while the other measures the height.

ABOVE The RouterRaizer can be used to make fine changes to the depth of an inverted router. Adjustments are made through the tabletop using a special wrench.

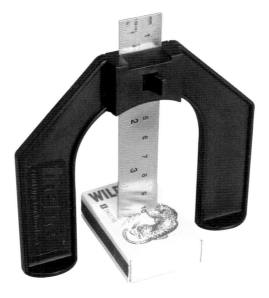

ABOVE The Depth Gauge is an invaluable aid for accurately setting cutter depths. The body is rigid plastic moulding with an opening 60mm wide. The steel rule has an 80mm measurement range and locks firmly in position with a cam lever. Various similar products are available, and it is possible to make your own version in the workshop.

The ultimate in smooth router raising is a RoutaLift, which looks like a standard router table but has a built-in height-adjustment mechanism. The router is suspended underneath and raised with a removable cranked handle from above. Easy to use, the RoutaLift is also extremely precise – one turn of the handle moves the router height by just 0.05in (1.3mm). The only disadvantage is that some measuring and drilling work is required to mount the router and to install the unit into the table.

The Trend Depth Gauge is very useful for depth adjustments and setting the fence on a router table. I would strongly recommend this product or a home-made version for every router user.

Guidebushes and bases

The guidebush is often said to be the most underused routing accessory and is certainly one of the most useful. If your router manufacturer provides no more than the standard issue, various major tool suppliers may well be able to oblige, although you may need to buy and use a sub-base to get the full benefit of the range of guidebushes available. American-style guidebushes are threaded and will need to locate into an appropriate base.

The Axminster Round Base is a universal Tufnol sub-base suitable for all routers. The integral guidebush will accept cutters up to 67mm, but if you require other sizes then the Round Base will accept a full range of American-style threaded guidebushes.

The Trend Unibase is also highly recommended. As its name suggests, the Unibase is an all-singing, all-dancing sub-base. When attached to your router you will be able to use a huge range of guidebushes. The only downside to this excellent attachment is that, at 8mm thick, you will be losing some cutting depth. The solution is to use extra-long cutters.

ABOVE In order to get the most of out the range of guidebushes available, it may be necessary to use a sub-base.

TECHNIQUE:

Using a Sub-base

TECHNIQUE:

Using a Sub-base

Whatever your router type, always fit the guidebush and/or the sub-base using a centring device. For 100% accurate guidebush work it is essential that the bush lines up with the spindle. I would always recommend using a guidebush-centring device, whatever the quality of your machine. If there is not enough 'play' in the guidebush fixing position or sub-base to adjust with a centring device, it may become necessary to re-drill some slightly oversized holes in the guidebush rim or to adjust the sub-base.

ABOVE The Unibase will fit virtually every router, with the only downside being the loss of cutting depth.

Another option, available through numerous US suppliers, is the Milescraft Turn-lock Baseplate, which is a 6in universal base made from transparent polycarbonate. This comes with its own guidebushes (which simply turn and lock into place), as well as an adaptor to fit normal brass or steel guidebushes.

Finally, an offset base is useful for added control when edge moulding. This can easily be made at home from MDF and a door handle.

Table routing

If you choose to make your own table, there is a wide range of ready-made products that for a price will make the job easier. Just some of the available kit is listed here, and you will need to check the appropriate supplier and manufacturer catalogues for the full story.

Push sticks are of course essential for table work and other machining – they give you a good grip and keep your hands safely away from the cutting edge. Work pads (which have a rubberized grip on their underside) are less widely used but offer excellent safety and control when using an inverted router. They are also particularly good for routing wide boards. Push blocks are simply a flat-handled pad featuring a step at the back that pushes against the workpiece.

NVR (No Volt Release) switches are another absolute must for home-made router tables. They provide safe and easy switching for machines of up to 1850W. Check with the supplier or manufacturer for details of any extra options that may be available.

Inserts are another essential part of table routing. The insert is recessed into the tabletop and has a hole through which the cutter protrudes – a set-up that provides solid support for the workpiece as it is moved into position. Although it is possible to make a table insert from MDF or ply, the ideal is to use the thinnest, strongest material possible. The plastic, phenolic resin, aluminium and steel used for specially made inserts make good options.

ABOVE A cranked spanner is a simple and effective way to overcome access problems when changing the cutter in a table-mounted (inverted) router. Similar versions are available from a range of suppliers.

General accessories

A successful router workshop will have all sorts of kit in addition to jigs that will help with the pursuit of routing. It pays to have a box full of nuts, bolts, washers, handles and anything else that might help you to construct a home-made jig. Accuracy is all-important for successful routing, so make sure that you always have a decent tape measure, some steel rules and an engineer's square or two to hand.

Some accessories are absolute must-haves; for example router mats, which are available from tool suppliers or DIY stores as anti-slip mats. Double-sided tape is essential for template routing – it can be used for temporarily fixing sub-bases, templates and a whole host of other accessories. Whilst I am sure that various manufacturers will insist that theirs is best, any brand of stationery-quality tape will do. Stronger industrial tape is normally too sticky, making temporary fixings too permanent.

FOCUS ON:

Vernier Callipers

FOCUS ON:

Vernier Callipers

Vernier callipers are very useful for precise woodworkers, particularly router users. The difference between a ½in and a 12mm cutter matters rather a lot, and is not always apparent once the shanks are worn.

I prefer using the non-dial variety because it impresses people and makes me look clever. Swiss-made versions offer the ultimate in quality, but cheaper models are fine for general woodworking.

ABOVE Vernier callipers can be used to introduce a high degree of precision to your work, which is particularly useful when routing.

ABOVE Useful routing accessories don't have to be complex or expensive. A router mat is a simple but essential accessory that provides safe, non-slip routing.

The corner chisel is a simple product that makes squaring out corner radii quick and easy. It is not recommended for top-class cabinetry, but it is an excellent product for the average woodworker.

As with most tools, if used with skill the Timber Repair Template or its basic principles can be used to great effect. A guidebush and collar are used with a template to recess and then refill with a damaged or unwanted piece of timber.

A profile gauge is a very useful tool on the odd occasion that it is used. Whether for transferring a profile to paper, or for a particularly awkward bit of scribing-in, the gauge makes easy work of a range

KEY POINT

If you're into making jigs, it would be well worth buying a selection of T-nuts, threaded inserts and so on – simply add a selection to your order next time you place one with a supplier. Also check catalogues for template materials and many other bits and pieces.

of troublesome processes. It is also an excellent tool for understanding complex moulding intersections.

Lastly, pocket whetstones and credit-card diamond stones are ideal for maintaining the keenest edge on your TCT cutters. The best all-purpose grade is a fine-surfaced stone of around 600 grit. For burr removal, honing cones are an option, although most woodworkers manage without and simply use the edge of a leather strop.

KEY POINT

Although keeping your table, fences and router bases clean should prevent friction problems, low-friction material (basically a hi-tech plastic) will reduce this down to a minimum. It is worth buying a roll of tape just to try it out.

KEY POINT

KEY POINT

1:5 Safety and dust extraction

The sharpness of the cutter, the high rotation speeds produced by the motor and the dust generated in use certainly gives a router the potential to be dangerous. However, if used properly and in accordance with a few simple guidelines, you are unlikely to suffer as much as a cut finger.

As well as learning how to handle the router safely, it as also important to wear the right kind of protective clothing and set up a dedicated working environment. A final issue to pay serious attention to is that of dust extraction. This isn't just a matter of keeping your workshop clean – prolonged exposure to the dust of some woods, especially manmade boards, can have serious health implications.

Workwear

When undertaking any sort of woodwork, your first thought should be for eye protection. Although in many instances a full-face visor is the best option when routing, an ordinary pair of safety glasses make it easy to abide by the basic safety rules that so many of us flaunt – especially if they are kept ready for action on top of the head. It is also a good idea to find a really comfortable pair that will not chafe your skin or give you a headache when worn for several hours.

ABOVE Glasses and ear defenders are essential pieces of safety equipment for routing. But don't just own them – use them!

Prolonged exposure to the noise of any size of router will damage your hearing, so a pair of ear defenders is another necessity. Various other earplug solutions are available. Apart from in the hottest weather, however, I normally stick to the standard variety because they are the easiest to get on and off.

A great many woodworking injuries are the result of heavy items such as routers, sheets of 25mm MDF and toolboxes being dropped on feet. I normally wear a pair of ancient steel toe-capped boots around the workshop, since I often need to handle large sheets (and can rest them over the toes) and because of distant and painful memories involving a Jack plane being dropped onto my foot.

I have also seen a plunged and locked router chew up a parquet floor between someone's feet. The near-victim was a virtual novice who switched on the plug of a router that was plunged and lying on its

KEY POINT

KEY POINT

Comfort is an important but sometimes overlooked part of workshop clothing and safety precautions. Remember: the more comfortable your safety gear, the more likely you are to wear it.

side on the bench. The big cutter and lack of a soft start meant that it kicked enough to fly off the bench – enough said.

Clothing

For many years, I only wore aprons at the bench. However, I have recently begun to favour wearing a turner's smock. This has several advantages: it fits over anything, can be sealed up to keep out even the highest-flying chips, and has back pockets that are excellent for holding useful items like tape measures and so on. I also have a cheap but serviceable welder's leather apron. This provides that extra layer of insurance when spindle moulding or routing, particularly when machining reclaimed timber.

⊙FOCUS ON:

The Workspace

If at all possible, have a dedicated routing bench – preferably one that enables you to walk around at least three sides and that is lower than a standard worksurface. I am 5ft 10in (1.78m), and so favour a routing surface of around 3ft (900mm) high. Another consideration is the provision of a lip around its edge to allow workpieces to be cramped firmly into place. I find that a 1⅛–2in (30–50mm) overhang is fine for basic cramping requirements.

The bench top should also have a false, sacrificial top that can be replaced once you have routed away the surface in certain areas. Being able to rout into the work surface makes a great many routing operations easier.

An extra-long hose suspended above the bench with a bungee cord is one of Ron Fox's many excellent ideas for improving your workspace. If you can run the hose extension to a fixed point on a wall, into which you can connect your extractor, you will have a perfect set-up. I would also recommend having a power point situated above the router bench, with the same sort of bungee set-up for the cable.

ABOVE Using a bungee cord to suspend a vacuum hose.

Dust

All fine dust is harmful, but this is particularly true of that produced by machining and sanding sheet materials such as MDF and particleboard. The combination of ultra-fine dust particles getting stuck in your lungs, plus the moderate to high levels of formaldehyde present in the glue, are enough to ruin anyone's health if exposed to them for long enough.

This is not a problem that can be ignored, so if you own and use a router you should also have an extractor. Routing produces lots of dust and woodchips, so apart from the damage to your lungs, unextracted waste material will make a terrible mess of your workshop!

Ultimately, your extraction and filtration systems can only work with what they are given, and the majority of dust thrown into the atmosphere will be down to the

good design or otherwise of your router. In-built extraction is still rare in routing and so most machines rely on a dust cowl and spout, which often work with varying degrees of success. However, whatever the system, having a decent extractor is a good start.

The amount of dust extracted away from the cutter also depends on the type of routing operation underway. Unless your parallel fence has an extraction port, edge moulding throws out the most dust because of the unconfined space under

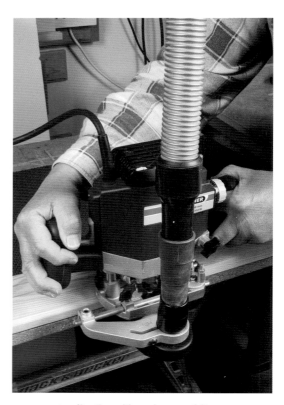

ABOVE Applications like edge moulding can throw out a lot of waste unless you have a specialized dust cowl.

KEY POINT

KEY POINT

Dust extraction is not just about keeping your workshop clean. Fine dust, especially that produced from sheet materials, presents a health hazard that must be addressed by using the correct equipment.

ABOVE Some guidebushes have been designed to aid the extraction of dust by allowing the air to be sucked through them.

half of the router. Guidebushes also give rise to problems – the extractor is trying to suck from above but is hindered by the bush, which blocks most of the air passage to the cutter.

Some guidebushes have been designed to counter this endemic problem, with large openings to allow the free flow of air. However, in the main the problem has never been properly addressed.

(FOCUS ON:

Dust Extractors

(FOCUS ON:

Dust Extractors

A purpose-built industrial vacuum extractor with automatic switching is the most efficient and easiest system for dust and chip removal. A domestic vacuum cleaner can provide a low level of extraction, but ultimately will not have anywhere near the amount of power or dust retention needed for machining hazardous materials like MDF. In a workshop there are two types of dust and chip extractors: High Pressure Low Volume (HPLV) and High Volume Low Pressure (HVLP). HPLV is suitable for routing, whereas HVLP extractors are for static machinery such as table saws and lathes.

ABOVE A purpose-built industrial vacuum extractor with automatic switching – the most efficient system for dust and chip removal.

Air filtration

Ambient air filters provide an equally important protective function. The ultra-fine dust particles that stay airborne for many minutes will contribute to the build-up in your lungs unless you wear a respirator or use an ambient air filter.

This additional piece of kit may seem excessive to some, but if you are moulding or rebating MDF, the volume of waste produced is very high. No matter how expensive your router or dust extractor, it is unlikely to help suck away all the dust produced.

ABOVE Battery-powered respirators, such as that shown top-left, are the best all-round option, particularly if you use a lot of MDF and other manmade boards.

ABOVE Ambient air filters provide an equally important protective function.

⒦KEY POINT

Whilst disposable cup masks offer enough protection in conjunction with vacuum extractors and/or air filters, they are not suitable as a stand-alone solution. Respirators are a better all-round option, particularly if you use a lot of MDF and other manmade boards. If you are working with no other protection or are particularly sensitive to dust, a self-contained respirator is the only solution.

⒦KEY POINT

1:6 Maintenance

A poorly maintained router will not work to its full potential and may, as a worst-case scenario, present a safety hazard. Taking good care of your tools will not only help to ensure that you get the best results possible, it will also make your kit last longer. Given the cost of some routing equipment, this is surely something that most of us cannot afford to ignore. Fortunately, with routing it couldn't be simpler – a few minutes' work every week cleaning and lubricating is all it will take to give you and your tools that professional aura.

All accessories and attachments also need to be properly maintained. As mentioned previously (see page 18), the collet should be checked for wear and all cutter shanks must be kept clean and free from rust. This will ensure that the collet holds the cutter shank firmly and allows it to rotate without any trace of wobble or vibration.

Router maintenance

Apart from regularly lubricating the plunge columns with a PTFE or silicon spray, an external spring clean and the brushing away of dust and debris using a dry brush or cloth should be all that is needed to keep your router in good working order.

Fine dust is the router's biggest enemy, and since the motor has to draw in air to cool itself, dust build-up on the windings will cause the most problems long term. The best way to deal with this is to send your machine for a major professional service every couple of years or so.

FOCUS ON:

Elu-style Switches

FOCUS ON:

Elu-style Switches

The old Elu-style of switch normally suffers the most from dust build-up, and even a reasonably well-maintained switch can develop a recurring problem. Since it is not possible to take them apart completely to clean them, it is often best to replace the whole switch assembly once this is the case. Consult the instructions before starting out and – it should go without saying – remember to disconnect the router's power supply before attempting even an initial inspection.

LEFT Changing motor brushes and the replacement of a faulty switch should really be the only internal maintenance jobs undertaken by router users.

Changing the motor brushes and the replacement of a faulty switch should really be the only internal maintenance jobs undertaken by router users. Anything else, such as replacing any worn bearings, should be the responsibility of a fully qualified service engineer.

Most router manufacturers worth their salt supply a spare set of motor brushes with the machine. If you don't have any spares, you should order some at your earliest convenience. This is the best way to guarantee getting the right parts, since by the time the first lot have worn down, your router model may be a museum piece.

TECHNIQUE:

Changing Brushes

Many router designs make it easy to replace brushes externally simply by removing a pair of plastic covers/holders from the top of the motor body. Other models will require part of the motor cover to be removed, but as long as you pay attention to how the electrics are wired up and don't touch the windings, this should not present the average woodworker with any problems. However, always remember to disconnect the power supply before attempting any internal work.

Bearing problems

Routers place a great deal of stress on ball bearings, and the ball-bearing races that hold the router spindle are prone to dust build-up and failure. Whatever the cause of worn or damaged bearings, the warning signs are quite obvious. To prevent further damage, any problems should be addressed as quickly as possible. Major repairs, such as replacing worn bearings, should always be carried out by a qualified service engineer, mainly because they require the partial dismantling of the router.

If your router motor becomes louder, emits different noises than usual or starts to generate excessive heat, it is most likely that solidified grease and/or damaged bearings are the fault. Another common symptom of bearing deterioration is noticeable vibration – but whatever your suspicions, the first thing you should ascertain is that you're not dealing with a collet or a cutter problem.

KEY POINT

General router maintenance is a simple and straightforward task. However, some parts, such as collets and ball bearings, are placed under extreme stress. These should be checked frequently, and faulty components repaired or replaced as quickly as possible.

TECHNIQUE:

Changing Brushes

KEY POINT

Cutter and collet maintenance

For safe, efficient routing it is essential that cutters and collets are maintained properly. TCT and HSS cutters can be kept perfectly sharp by honing the edges on a pocket diamond stone, and all cutters can and should be cleaned of resin and dust build-up using lighter fluid or contact adhesive remover (but remove the bearing guides first). After cleaning, cutters should be sprayed with PTFE in order to reduce further build-up.

The collet should also receive some care and attention with cleaning fluid, and a

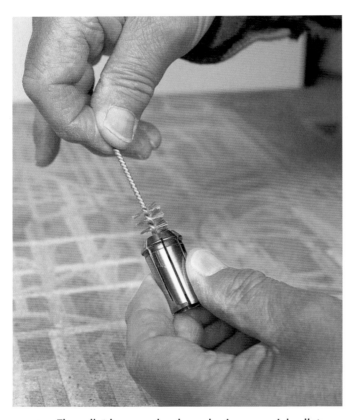

ABOVE The collet bore can be cleaned using a special collet brush and some cleaning fluid.

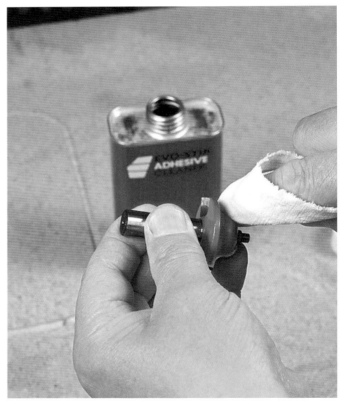

ABOVE All cutters should be cleaned of resin and dust build-up using lighter fluid or contact adhesive remover.

brass brush can be used to clean away any rust or other deposits. In addition, the collet should be checked frequently for wear or damage. Ordinary wear generally takes the form of 'bell-mouthing' – where the top and bottom of the collet become wider than the central portion – which can lead to slipping and/or wobbling. As the tolerances are measured in tiny amounts, such wear may not be visible to the naked eye. However, the symptoms include excessive vibration and collet markings or rings on the shanks of cutters. If there are any signs like this, the collet should be replaced immediately.

ABOVE TCT and HSS cutters can be kept perfectly sharp by honing the edges on a pocket diamond stone.

ABOVE After cleaning, cutters should be sprayed with PTFE in order to reduce further build-up.

Part 2:
Techniques

2:1 Get routing

Once you've bought a router, you will naturally be keen to use it.
First of all: be safe. Read the background information and general
theory, especially the section on safety (see page 71), before attempting
what comes next. Developing a good understanding of the basics is
fundamental to the eventual success of your endeavours, and therefore
an essential part of your enjoyment.

You should also think carefully about what it is that you want to make
before leaping in. This may sound obvious, but working on a project is
much better than practising solely on offcuts – if the workpiece matters,
you are likely to take more care over it and yourself. Also, your router
can be used for a huge range of applications and it is important to be
clear about the techniques and equipment required for the job in hand.

In control

A router and its cutter can be guided in eight different ways, although the number of applications to which they can be applied are endless. The following headings list the different methods, explaining each and raising some key considerations.

Bearing-guided cutters

These 'self-guiding' cutters are able to follow the edge of a workpiece or template by the use of a rotating bearing guide. This can be located above or below the blades. For flush trimmers and profiles the bearing guide is of the same diameter as the smallest part of the cutter. For rebating sets, on the other hand, the offset between the bearing radius and the cutter radius equals the width of the rebate.

ABOVE Bearing-guided routing. In this case, the bearing is located between the shank and the blades.

Bearings can also provide a useful secondary function when used in conjunction with profile cutters: bearings of varying sizes can be used to produce different and, most importantly, repeatable depths of cut.

Once you have sent your cutters to a saw doctor to have the edges re-ground, there will be a slight discrepancy between the width of the bearing and the cutter diameter. This is something that you will need to account for in your work, although it is a particular problem with trimmers since they will no longer trim completely flush.

Guidebush and template

A guidebush is a metal ring that fits into the base of your router. The inside edge of the guidebush features a collar, and this runs against an edge, which can be straight or irregularly shaped (such as that created by a template). The cutter follows this path to whatever depth it is set.

ABOVE Guidebush routing. Used in conjunction with a template, this is an excellent and flexible method for guiding a router.

The use of guidebushes in conjunction with templates will be discussed in greater detail as part of chapter 2:2 (see page 97). They key phrase to remember at this stage is 'offset'. The outside edge of the guidebush collar will always be of a greater diameter than the cutter, which means that the template is a parallel offset of the router path.

Guide rails

A guide rail is an advanced extruded aluminium straight edge, into which is fixed a connected baseplate. The router slides along this guide, and the result is almost foolproof routing. However, as usual, the results are only as good as your setting up.

Straight edges and clamp guides

A router's base (which usually has rounded and straight sides) can simply be run against a straight or irregularly-shaped edge in order to guide the cutter to groove, house, cut or trim a workpiece.

A straight edge can be provided by anything from an MDF offcut to an extruded aluminium clamp guide. I find it useful to keep lengths of ply and MDF boards for this purpose. The already-straightened outside-edge offcuts from 8 × 4ft (2.4 × 1.2m) sheets tend to be the most suitable, since you will rarely need to rout anything longer than 8ft (2.4m).

ABOVE Routing against a straight edge, which has been securely clamped to the workpiece.

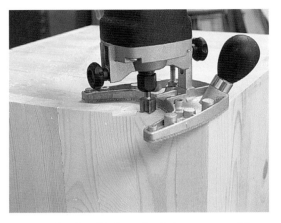

ABOVE Hand-guided routing is best reserved for simple tasks like flush cutting.

Hand guided

Whether flush-trimming (as above) or with a base extension, using the router freestyle can require a little more care than usual. This is because your fingers are more likely to be holding onto the base and will therefore be nearer the cutter.

Parallel fence

This piece of kit is supplied as standard with all routers. It will enable you to add rebates and mouldings to straight edges using unguided cutters and to rout slots parallel to edges. The plastic facings on the fence can usually be adjusted to reduce the gap around the cutter. By so doing, the path of the router and the fence as you lead onto and off of the workpiece is made smoother.

Fine fence adjusters are available with some of the more expensive amateur routers and many professional machines. Of course, they are not essential, and many of us managed for years without them. However, while not vital, even the crudest of adjusters can really make a difference to the speed and ease of your set-up.

LEFT An example of using a parallel fence to guide the router.

Pin-guided cutters

The pin at the base of the cutter works in the same way as a bearing guide. The only differences are that it rotates with the cutter and that it is able to fit into smaller spaces. Care is needed when using pin-guided cutters, since the friction of the pin has a tendency to cause burning and to 'bite in'. This can be a particular problem if it is not kept clean from resin and the build-up of other unwanted substances.

Roller guide

A roller guide can be used to facilitate edge moulding with a non-self-guided bit by means of a roller that follows an uneven edge. This awkward piece of equipment gives inconsistent results and so my advice is to use a roller guide only if you have to.

ABOVE A pin-guided cutter works in much the same way as a bearing-guided cutter, but care must be taken to avoid burning.

LEFT A roller guide attached to a router. This can produce inaccurate results and is not my first-choice guiding method.

Trammels, circle and ellipse guides

Circles and arcs are most easily routed using a simple beam trammel, and this will be suitable for the majority of applications. For circles with a diameter of less than 100mm, however, a manufactured or home-made small circle guide will be necessary. Extra-long router rods are available for larger radii, and after that you will need to make your own trammel or use a large ellipse guide in circle mode. Ellipses can be routed using manufactured or home-made jigs. However, great care must be taken when making your own jig, as any inaccuracies in the edge will be magnified when cutting.

KEY POINT

Remember that your finished piece can only be as good as will be allowed by the materials you use in its construction. For example, any irregularities along the edge of a template will be mirrored by any edge-following method.

ABOVE A basic trammel set-up makes routing circles and arcs easy.

Direction of cut

Whether freehand routing or using a table set-up, it is vitally important to choose the correct direction for machining. When routing by freehand, it doesn't matter whether you push or pull the router along the workpiece – it's best to find out what works best for you. However, it is essential that you present the cutter in the right way. The diagram, right, illustrates this point. Whichever way you are looking at

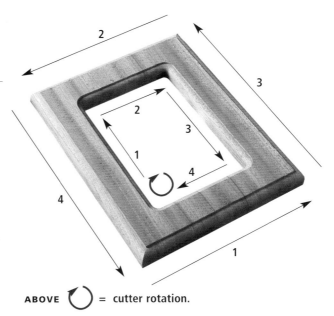

ABOVE = cutter rotation.

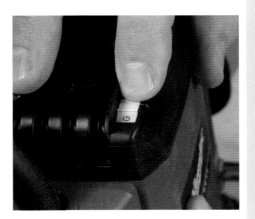

(FOCUS ON:

Variable-speed Routers

Variable-speed routers allow you to select the most effective speed for the cutter. This should take into account your machine's motor size, the diameter of the cutter, the type of material being cut and the feed rate being used. You should always refer to the manufacturer's instructions for the precise cutter speeds, but the following table gives a rough guide. The information listed applies to softwood and manmade boards. When routing hardwoods, rather than deviating from these guidelines, you should merely reduce your feed rate.

Diameter	Speed
3–25mm	22–24,000rpm
25–50mm	18,000rpm
50–72mm	16,000rpm
72–89mm	8–12,000rpm

ABOVE Selecting the correct speed is easy. In this case, the thumbwheel is marked with letters that correspond to different speeds (the router's manual will contain a chart allowing you to translate each letter to a speed).

(FOCUS ON:

Variable-speed
Routers

your workpiece, you should always be moving the router counterclockwise around the outside. For routing the inside of, for example, a frame, the movement should be in the opposite direction – i.e. clockwise.

Of course, when a router is inverted in a table everything is turned onto its head, but if you remember to cut only into the oncoming rotation of the cutter you will be safe. If in doubt, stop the machine and get things sorted in your mind.

When routing enclosed spaces where there is timber all around the cutter, the direction of your pass doesn't matter because the cutter is restrained in every direction. However the moment the cutter is exposed on one side the normal rules apply.

Equipment limitations and guidelines

To prevent too much stress being placed on the cutter, nearly all routing jobs will require multiple passes rather than a single cut. Trend's guidelines on the subject of cutting depth are about as straightforward as you can get: 'the depth of cut should not exceed the diameter of the cutter when using cutters of up to ½in diameter, or the diameter of the cutter shank, whichever is smaller.' Basically, this means that a ¼in cutter should not be plunged more than 6mm at a time, and a ½in cutter should not be plunged more than 10mm.

ABOVE After zeroing the cutter on the workpiece, the depth is set using the diameter of a drill bit. This can then be increased as the job progresses.

KEY POINT

'Burning' or 'scorching' can be a problem when routing. It can be caused by several factors. It may be a result of the cutter being fed to slowly, of lingering on a point during routing, or of cutter bluntness.

Feed rate

Although there are suggested optimum feed rates (in feet or metres per minute) for cutters, the best way of judging the feed rate is to listen to the motor and to look at the results. Providing you follow the depth-of-cut guidelines, the motor should not reduce more than a small amount in speed, and there should be no scorching of the workpiece. Providing your cutter is sharp, any scorching that does occur will be the result of routing too slowly. Using the optimum feed rate will produce better cuts and increase the life of the cutter.

FOCUS ON:

Grain Direction

FOCUS ON:

Grain Direction

The direction of your routing in relation to the direction of the wood's grain can be significant. Routing with the grain should present few problems because you are not cutting through the fibres of the timber. However, more care needs to be taken when routing across the grain, as 'spelching' or 'break-out' can occur at the lead-off corner.

This can be prevented or removed using various methods. If you are moulding or rebating a cross-grain edge but don't intend to also rout the edges, you can use sacrificial battens to protect both the lead-in and lead-off corners. The latter batten is obviously the most important, but the lead-in will give your edge an extra crispness.

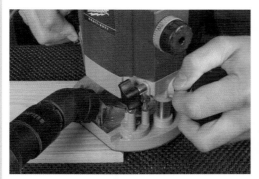

ABOVE Start by routing the end grain so that any breakout is machined away when routing the length.

ABOVE Repeating the process at the opposite end of the workpiece.

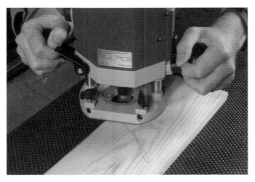

ABOVE Routing along the grain.

ABOVE Repeating the process on the opposite side.

2:2 Template-guided routing

Templates are simply routing patterns around or within which a cutter follows. They can be bought commercially, although most are workshop-made from wood or plastic for specific and general use. Essentially, a template is a quick and easy way to duplicate any given shape with great accuracy and consistency – saving time when routing multiples of the same part, or perhaps enabling two irregular-edged boards to be edge-joined with great precision.

At the simplest level, template-guided routing can be achieved using the router base itself. However, this method obviously imposes severe limitations on the levels of accuracy and detail that can be achieved. Using either a bearing-guided cutter (which is particularly useful for inverted routing) or a guidebush (which is ideal for freehand work) provides much greater flexibility.

Whichever method is used, routing with templates gives even the most modestly skilled woodworker the chance to produce accurate, fast and repeatable results.

Bearing-guided cutters

Bearing-guided cutters are able to follow the edge of a template through the use of a rotating bearing. This bearing is the same diameter as the cutter blades at their narrowest point and so will duplicate the template exactly. Although it is entirely appropriate to use this cutter and template combination when routing freehand, the technique is particularly well suited to table routing.

KEY POINT

When you are making a template, don't forget that you can rout internal radii only as small as your cutter/bearing diameter. Anything smaller and you will have to resort to a suitable cutter, guidebush and offset template, or possibly a pin-guided cutter.

FOCUS ON:

Bearing-guided
Routing

FOCUS ON:

Bearing-guided Routing

Templates for bearing-guided routing can be much more straightforward than for guidebush routing simply because there are no offsets to consider – the template will produce exact copies of itself. A square-cornered frame simply requires a square template; if, however, you require corner radii, these can either be machined using a scrollsaw, bandsaw, jigsaw, by sanding or better still, by using a router. To rout the corner radii most precisely you can use a router fitted with a guidebush, although the external radii possible will be limited to the radii of your guidebushes. For a step-by-step guide to routing a picture frame with a bearing-guided cutter, see page 102.

98

Most straight and profiled bearing-guided cutters have bottom (at the opposite end to the shank) rather than top-mounted bearings. In many ways, straight cutters with top bearings are better suited to freehand template routing – the template can be placed on top of the workpiece, often resulting in an easier set-up and a clearer view of the job in hand. However, if you only have the bottom-bearing variety and don't have the funds for top-bearing cutters, all is not lost. With good preparation of the workpiece (a more-or-less consistent width of waste around the edges) and by marking potential trouble spots on the upper side of the workpiece, you should be able to cope admirably with the template being underneath.

If you choose the bearing-guided approach for table routing (see page 111), remember that small and narrow components are often best routed on a table with a lead-in pin or a batten and overhead guard. I recommend routing large components, such as chair legs, on a decent-sized table with a bearing-guided cutter, since it is easier to control the cut (partly because of the improved visibility). Don't forget to present the workpiece on the correct side of the cutter (so that the blades are turning against the direction of feed), and be particularly careful with end-grain and sides shorter than about 2in (50mm).

Guidebushes

The outside edge of a guidebush collar is designed to run along the edge of a template or jig. The only prerequisite is that the material is thicker than the guidebush collar is deep. A cutter protrudes through this collar and follows the outside of the collar at an offset distance. The exact offset is related to the distance between the outside edge of the cutter and the outside edge of the guidebush, and can be calculated using the equation shown in the panel, right.

A guidebush is appropriate for routing within templates and jigs to produce housings, panel moulds, hinge recesses, inlayed repairs, dovetails, and so on.

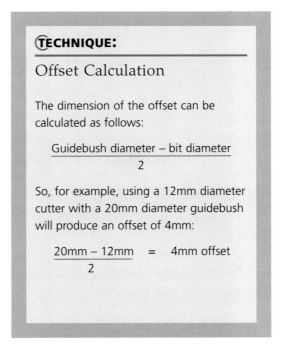

TECHNIQUE:

Offset Calculation

The dimension of the offset can be calculated as follows:

$$\frac{\text{Guidebush diameter} - \text{bit diameter}}{2}$$

So, for example, using a 12mm diameter cutter with a 20mm diameter guidebush will produce an offset of 4mm:

$$\frac{20\text{mm} - 12\text{mm}}{2} = 4\text{mm offset}$$

TECHNIQUE:

Offset Calculation

ⓕFOCUS ON:

Guidebush
Routing

ⓕFOCUS ON:

Guidebush Routing

Producing a frame using a template and guidebush is an excellent and productive way of learning the basic principles of guidebush routing. The frame made in this example (see page 105) has corners of the same external radii as the guidebush used. If you want to achieve square external corners using a template you'll need to use a bearing-guided cutter. Of course, the internal radii of all template-routed corners the same dimension as the cutter.

ⓚKEY POINT

ⓚKEY POINT

When routing within a workpiece, remember to rout in a clockwise direction. When routing around its outside edge, the feed direction should be the exact opposite: counterclockwise.

One of the chief advantages of guidebushes is that it is possible to use different cutters and guidebushes with the same template.

Many commercial templates and jigs are available for use in conjunction with guidebushes. These include comb, dovetail, hinge, letterbox, mortise, lock, mortise-and-tenon, stairway, timber-repair kits, veneer-inlay kits, and worktop jigs. All require specific guidebush and cutter combinations. Further descriptions of jigs can be found in chapter 2:4 (see page 121).

In order to make your own template or jig for use with a guidebush, you must first decide on the diameter of cutter and guidebush you will use, remembering that the diameter of the guidebush will always be greater than that of the cutter. However, since the smallest guidebush diameter available is 7.74mm, your templates' internal radii or apertures will be restricted to this dimension.

Template technical

For the home user, 6mm MDF is normally the most appropriate material for making templates, although where guidebushes are concerned it simply needs to be thicker than the depth of the flange. Remember that any imperfections in the template will be transferred to your work, so make every effort to get it right. If necessary, dents can be successfully repaired with car body filler! Never work with anything other than a perfect template.

Where possible, use a router to make your templates. Utilizing a corner-radius template (home-made or otherwise) and straight edges can make for a time

KEY POINT

Double-sided or carpet tape can be used to stick your template to a workpiece, but don't use too much or it might be almost impossible to separate the components without damaging them. To be on the safe side, experiment with a test piece first.

consuming job, but will ultimately produce the cleanest results. A combination of files, rasps, sandpaper and a bobbin sander (or a sanding drum in a drill press) can help shape your templates, particularly if you have used saws to cut them out.

Once you have made your template, you will need to transfer it to the workpiece to either mark out and trim or attach and trim it to within 2–4mm. The less waste the better, but don't risk the workpiece to save a fraction of an inch of waste.

Whether you use regular double-sided or carpet tape, sticking your template to a work-piece is normally the best option. Having said that, panel pins can be used as a quicker solution where appropriate, and have the further advantage of being easier to remove.

You will be able to reuse double-sided tape, but only if you apply it firmly to the template first, then stick this gently to the workpiece. You might be able to get a couple of adhesions from regular double-sided, and several from carpet tape. I use a large (although not sharp) kitchen knife to separate the components gently. However, if you don't want to reuse the tape on a template, you can where possible separate the assembly with an opposing twist, which will help to tear off the tape and make it easier to remove altogether. Tape residue can be removed using lighter fluid or, a little less easily, with mineral spirits (white spirits).

KEY POINT

Bearing-guided routing

1 The first step is to make the template. Cut a 6mm MDF blank with the external dimensions of your required frame, minus the offset produced by your guidebush and cutter combination. Remember that the offset needs to be applied to all four edges. Here, a 16mm guidebush and a ¼in cutter produces an offset of 5mm. Stick or pin the blank to an oversize 6mm MDF board and fix this assembly to a sacrificial worktop.

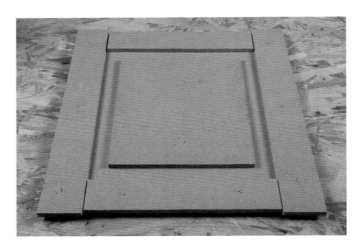

2 Stick or pin support strips around the outside of the blank, leaving a gap sufficient for your guidebush. You are now ready to rout the outside of what will become the template.

3 Rout counterclockwise around the blank in two depth passes. You will be left with a straight-sided board that has perfectly rounded external corners. Detach and discard the remains of the blank and, if necessary, re-stick the workpiece you have cut.

4 Mark out the central aperture on the workpiece and then stick or pin supporting guides at the 5mm offset outside the cutout line.

6 Attach the template to the frame material, which should be oversized by 4–6mm in each direction. Mark out the central aperture and cut out the waste (leaving 2-3mm spare inside the line) using a jigsaw or scrollsaw.

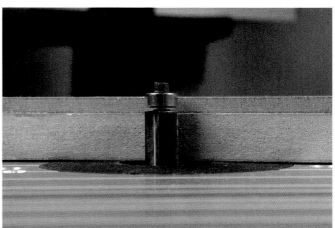

5 Plunge and rout clockwise in two depth passes around the inside of the guides. You will be left with a perfectly formed template for bearing-guided routing.

7 Set up a router table as shown, with a bearing-guided straight cutter. Set this to a height sufficient to remove the waste in one pass.

Bearing-guided routing

8 Make sure that you present the workpiece to the cutter correctly. When trimming the outside of the frame, you'll be passing the assembly counterclockwise against the rotation of the cutter. For the interior, you'll need to pass it clockwise against the rotation of the cutter. Be careful when routing such a small object – always hold the workpiece firmly and decide what you are going to do before you start.

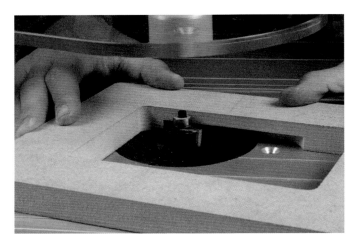

9 Once completed, remove the template and replace the cutter with a bearing-guided rebate cutter. Rout a rebate in an appropriate number of passes, once again remembering to present the workpiece to the cutter correctly.

10 Replace the cutter for a bearing-guided profile, in this case a 45° bevel. Flip the workpiece upside down and rout the internal decoration. Add an external profile if required.

Guidebush routing

1 The first step is to make the template. Cut a 6mm MDF blank with the external dimensions of your required frame, minus the offset produced by your guidebush and cutter combination. Remember that the offset needs to be applied to all four edges. Here, a 16mm guidebush and a ¼in cutter produces an offset of 5mm.

2 Stick or pin the blank to an oversize 6mm MDF board and fix this assembly to a sacrificial worktop. Then, stick or pin support strips around the outside of the blank, leaving a gap sufficient for your guidebush. Rout counterclockwise around the blank in two depth passes.

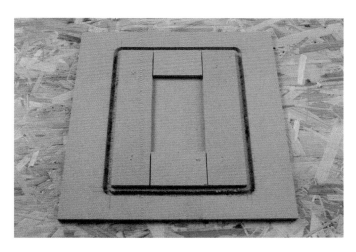

3 Mark out the internal opening, again taking into account the offset. This time, however, the offset should be added to the width and height of the opening (instead of subtracted, as before).

Guidebush routing

4 Pin or tape four 6mm guide strips to the marked-up lines of the internal offset.

5 Tape or pin a 6mm central support to stop the router from tipping over.

6 Rout clockwise around the inside of the battens until you have cut through to the table top.

7 Remove the completed template from the set-up.

8 The next step is to rout the frame. Attach your template to an oversized workpiece and then secure this assembly to a sacrificial worktop. Attach the external support strips. These should be placed in excess of 16mm from the outside of the template in order to create a large enough channel for the 16mm guidebush.

9 Using a ¼in straight cutter, rout counterclockwise around the exterior perimeter of the template in several depth passes until you have cut right through to the worktop.

Guidebush routing

10 Now add a central support within the frame, leaving a channel large enough for a 30mm guidebush.

11 Replace the 16mm guidebush with a 30mm one, and rout clockwise in several passes around the interior perimeter until you have cut right through to the worktop.

12 Still using the 30mm guidebush, fit a 19 or 25mm straight cutter and then rout to a suitable depth for the picture rebate using the same interior perimeter.

13 Remove the now isolated assembly and detach the template. Flip the frame over and stick it more or less in the same position.

14 Using a bearing-guided profile cutter or as in this case a 45° bevel, rout the interior perimeter on the face of the frame.

15 Add an external profile if required.

2:3 Table routing

Routers are generally thought of as hand-held tools, but that isn't the only option. Not that table routing is massively different – the router is simply inverted and mounted under a stationary surface. The main upshot of this is that the workpiece is brought to the cutter, rather than the other way around.

Table routing solves many of the problems presented by routing freehand, simply because it becomes easier to control the workpiece. Edge moulding, template following, joint cutting and many other applications are well suited to the table. If you have purchased a router and become enthused by its possibilities, table routing is the way forward.

Table anatomy

Although tables are designed a little differently, they all adhere to the same basic principles. The following list gives a rundown of the main requirements you'll have of a router table. Certain features are luxuries, but most, such as hold-downs or guards, really are necessities:

- a flat, rigid table top
- sturdy frame or bench fixing
- a vertical and a horizontal hold-down
- dust extraction facility in the fence
- a guard to keep the fingers away from the cutter
- mitre fence/sliding carriage
- NVR switch
- adjustable lead-off fence (a rare luxury)
- a router raising device.

Many pros prefer a cast or extruded metal table top. Whether yours is MDF, Phenolic resin or aluminium, there will be very little difference in performance as long as it is flat, stable and smooth.

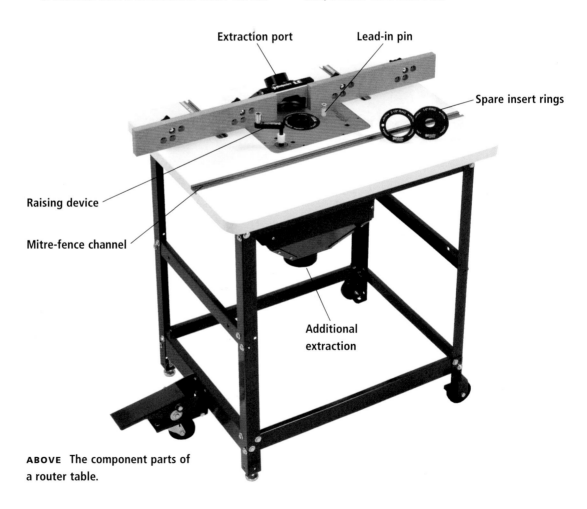

Extraction port

Lead-in pin

Spare insert rings

Raising device

Mitre-fence channel

Additional extraction

ABOVE The component parts of a router table.

Dust extraction through the fence is the best and most common solution, and is improved by having sliding fence facings that can be adjusted to minimize the aperture around the cutter. This helps to concentrate the suction, which in turn reduces the amount of dust thrown up into the air.

Mitre fences rarely have an accurately set up scale or a guide bar that fits snugly into the channel, and this typically produces a degree or two of 'waggle'. However, once you get used to your machine's peculiarities, this device aids all sorts of operations. I would recommend

constructing a sliding carriage (see the routing-table project on page 151) for all of those 90° passes that are difficult to keep square with a mitre fence.

An NVR (No Volt Release) switch is an absolute must for router tables, providing safe switching at your fingertips. An NVR enables the router's switch to be left in the 'ON' position and then operated from a convenient position on the outside of the table.

Whether you use a fine height adjuster or a device such as JessEm's Rout-R-Lift, the ability to raise and change cutters is the key to successful table routing. Always check that a raising solution provided with a table is compatible with your intended router.

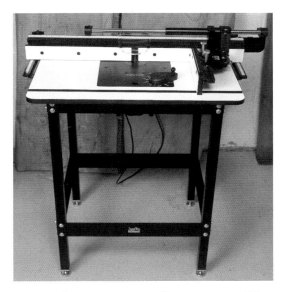

ABOVE Professional-quality kit comes at a price, but offers high build quality and a host of useful features. This CMT routing table incorporates a heavy-duty fence and a cabinet base for storage.

ABOVE The JessEm Rout-R-Lift system – possibly the ultimate router-table option.

(FOCUS ON:

Which Router?

(FOCUS ON:

Which Router?

The ideal routing set-up is to have two or more routers – a ¼in or ½in router for the bench and a ½in machine that spends most of its life inverted in a table. This is by far the best option, providing that the table-mounted machine has the option of a ¼in collet; that way you can still get the benefit from your full range of cutters. The ideal table router will have at least 1200W, although 1850W plus will give you access to all sizes of cutter. Of course, it is also useful to have a small bench-top router table such as that made on page 154. This is designed for use with an 850W machine, although with a couple of dimension modifications you could use it with anything up to 1200W.

LEFT The Freud FT2000E is an affordable professional-quality router. The plunge base, long collet and micro-adjustment depth control mean that it is well suited to inverting.

RIGHT The CMT 1850W router, which is based on the DeWalt 625EK. This is a classic pro workhorse and is excellent for table use.

Although it is relatively easy to make your own table, many beginners understandably opt for a ready-made set-up. However, for those on a tight budget, the home-made option is certainly a good way to get what you need for less.

Table-routing applications

By giving the user increased control over the workpiece, a table-mounted router is particularly well suited to certain routing applications. A few of these are outlined below, along with some suggestions for getting the most out of each technique.

Profiling

When using a fence to edge-mould, the cutter should never be more than half exposed and the workpiece should certainly never be passed between the cutter and the fence. Decent hold-downs are vital to achieving good results, although some low-cost routing set-ups come without any. The most common set-up is two vertical (a lead-in and a lead-off) hold-downs with a single, central, horizontal pressure board. Featherboards can be made easily from MDF or alternatively bought as accessories from any major supplier.

When profiling the whole or part of an edge it can become necessary to support the workpiece on the lead-off fence

ABOVE Routing a profile. Here, the workpiece is being supported by two featherboards (one horizontal and one vertical), which help to prevent it from vibrating against the cutter.

TECHNIQUE:

Workpiece
Support

TECHNIQUE:

Workpiece Support

One of the key factors in successful table routing is workpiece support. This can be improved in several ways. One method is to introduce a continuous fence, which is planted on the front of the two standard fences. It needs to be attached and then gradually pulled back over the cutter in order to expose just the right amount of profile for the job.

You can also improve matters by making the workpiece itself more rigid. Whenever possible, rout a decorative mould such as a beading onto the edge of a wider board and then rip this off using a table or a bandsaw. The board edge will then need to be run over the surface planer before the process is repeated again.

If you have already prepared your timber to a small section, you simply need to get the pressure of the hold-downs exactly right so as to minimize any movement or vibration of the workpiece against the cutter. You can also make a tunnel jig that will keep the workpiece completely controlled.

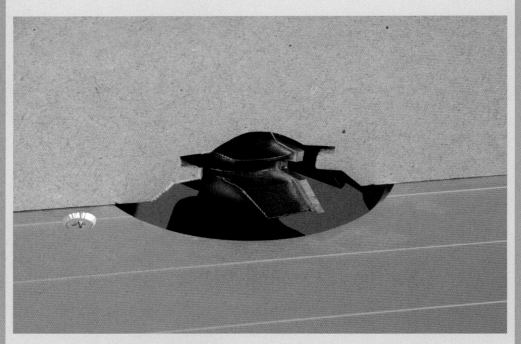

ABOVE A continuous fence and a lock mitre cutter. Part of the cutter's profile has been exposed, but only the correct amount required to perform the job in hand.

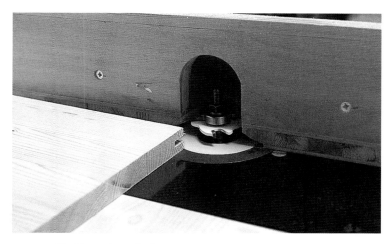

ABOVE Slotting a board edge.

because the face has been machined away. This is easily achieved by adding a suitable support strip using double-sided tape. Sand or plane a small chamfer at the strip's leading edge to aid the workpiece's smooth passage. Some advanced tables have an adjustable lead-off fence, similar to those found on spindle moulders, which can be adjusted to provide such support. However, these can still need support strips.

Slots, grooves and rebates

A table set-up is ideal for the fault-free routing of biscuit slots, panel grooves and, of course, rebates. The set-up for such operations is quite straightforward, but it always pays to try some test cuts to check the height and/or depth of the cut. Take note of advice on cutter depth and feed speed (see page 94) when table routing, since the temptation is often to machine off too much too quickly.

Bearing-guided routing

As shown earlier (see page 88), bearing-guided trimmers or profilers are able to follow the edge of a template by the use of a rotating bearing (which will usually be of the same diameter as the cutter blades at their narrowest diameter). Although it is entirely appropriate to use this cutter and template combination when routing freehand, the technique is particularly well suited to table routing – small components

---

KEY POINT

If you need to straighten or clean up the edge of a piece of timber, pin or tape a straight-edged template to the workpiece and run this against a bearing-guided cutter. You may need to do this in stages if the waste for removal is more than is reasonable for one pass.

KEY POINT

are more easily and safely handled on a table, as are larger workpieces such as furniture parts. Whatever you are routing with a bearing-guided cutter, for safety you should always utilize a lead-in pin or batten and some sort of guard.

Stopped routing

It is possible to 'drop' the workpiece onto the cutter and then move it between stops to achieve mortices, stopped housings and stopped chamfers. However, this method can be problematic and dangerous because it is often difficult to hold on to the workpiece. I would therefore advise against attempting anything apart from stopped chamfers, at least until you are an experienced table user.

ABOVE Using a bearing-guided cutter on a table. Note the guard, extraction and lead-in pin.

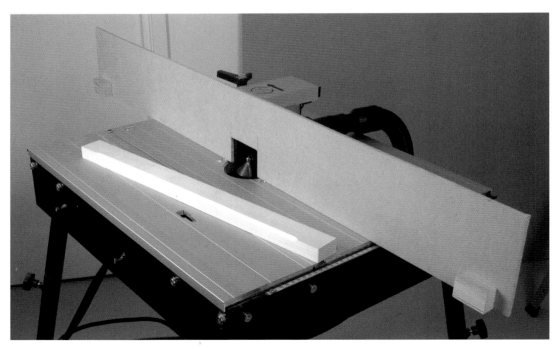

ABOVE Stopped chamfers are possible by attaching an additional fence facing with end stop blocks.

ABOVE Care and push-sticks are needed when the chamfers stop close to the end

ABOVE Edge planing. Straightening a length of timber using a template.

ABOVE Don't be tempted to remove too much at a time when edge planing.

2:4 Commercial jigs

To get the most out of your router you will need to be able to source the right jigs. There is a huge range of jigs to suit every conceivable task, and, as we shall see in part three (see page 149), many of these can be made in the workshop with little trouble or expense.

In terms of commercially available jigs, the most valuable are the dovetail jig and mortise and tenon jig. Dovetails and mortise and tenon joints – once the epitome of fine craftsmanship, and worthy time-consuming techniques – can be cut both quickly and accurately, with results comparable to handcrafted work.

In either case, different grades of equipment are available and it is important to get the right jig for your individual needs. This chapter offers an overview of the products available and advice on choosing the right one.

Dovetail jigs

There are two main types of dovetail joint: half-blind joints and through joints. Both are formed by interlocking a set of pins with a set of tails; a half-blind joint differs in that the cut does not go all of the way through the board and the ends of the dovetail are concealed on one side. The ability to hand-cut a dovetail joint has long been considered the measure of a craftsman, but now that modern glues are so efficient, dovetail joints are, despite their inherent strength, mostly decorative details cut using jigs.

Until a few years ago, it was possible to spot a jig-cut joint from a mile off. However, developments have raised the level of sophistication of dovetail jigs to the extent that they can now produce an impressive array of complex and decorative

ABOVE Examples of half-blind dovetails for the fronts of drawers.

ABOVE A selection of 'through' dovetail joints.

joints for the average woodworker. All that is needed is a bulging wallet and a methodical mind. There are, however, several basic jigs on the market that can produce straightforward results. These are good enough for most woodworkers' ambitions and won't break the bank.

Fixed-comb jigs are the most basic set-ups and will only cut one size and pitch of half-blind dovetails. The fixed size of these templates also means that the width of your board is governed by the multiple widths of each tail, and the thickness of the board is also limited.

LEFT Three fixed-comb dovetail jigs. These are dovetail jigs in their most basic form.

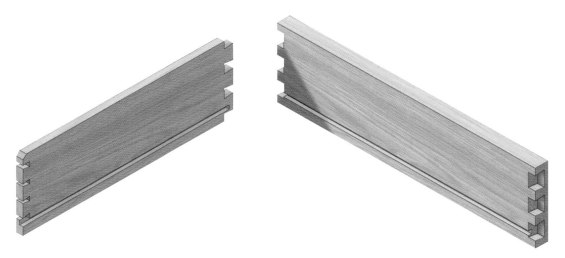

ABOVE A drawer side (left) and front (right) showing both types of dovetail joints. The front has half-blind sockets, whilst the side has both half-blind and through-joint tails.

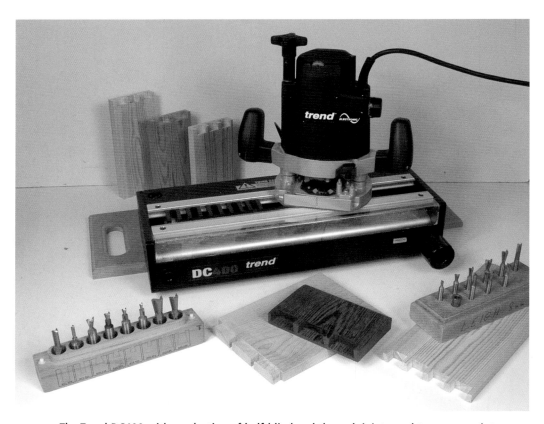

ABOVE The Trend DC400 with a selection of half-blind and through joints, and two appropriate sets of cutters.

ABOVE **The chest was made with a Leigh jig, the left box with the Trend DC400 and the right hand box with a WoodRat.**

KEY POINT

The more control and options you have, the more you pay. Top-range jigs can be up to eight times more expensive than their simpler counterparts. However, the difference is clear – you are either buying a straightforward jointer or a design tool.

More sophisticated jigs such as the Leigh, WoodRat or Trend DC400 can cut a variety of widths, pitches and board thickness to produce an infinite number of variations. By doing so, they give back the creative control to the woodworker that hand-cut dovetails once offered.

Specialist jigs

The Trend Mortise and Tenon Jig, the WoodRat and the Leigh FMT couldn't be more different. However, each has particular talents and is suited to a different pocket. The Trend is the cheapest of the three but still offers a great range of joints. Setting up is easy, with the same router-locating aperture being used to rout both the mortise and tenon. The mortise is cut in stages by the same diameter cutter, using a guidebush that runs against the outside edge of the aperture. A collar is then added to the guidebush in order to lock its movement in one direction so that it cuts the mortise. Joints can be angled from –10° to +45° and are as deep as your cutter and plunge will allow.

The WoodRat is extremely versatile. This highly original jig is able to cut lap, finger, housing, mortise and tenon joints, plus dovetails and profiles. The workpiece can be clamped and then moved with great accuracy around or past the cutter, which (uniquely) is done in the same direction as the cutter rotation. This is entirely safe because the workpiece and router are cramped securely, and has the effect of giving superior quality cuts in end grain.

The Leigh FMT is without doubt the ultimate mortise and tenon jig and is fast becoming the preferred option for cabinetmakers. With up to 20 different

KEY POINT

(FOCUS ON:

Choosing a
Dovetail Jig

(FOCUS ON:

Choosing a Dovetail Jig

There are two types of dovetail joint – 'through' and 'half-blind' joints – and several different types of dovetail jig. Fixed-comb jigs are the most basic since they are limited by their single spacing comb and can cut only the half-blind variety at one size and pitch. Because of their lack of adjustment, the board widths that you use with the most basic of these jigs (300 or 600mm wide) must be of exactly the right dimension to give equally sized pins at either end. Some slightly more advanced jigs have adjustable stops at either end of the comb so that you can centre the board and vary the width of the end cuts. These will still cut only half-blind dovetails, but at least give some control back to the user.

Mid-range jigs can offer this adjustable end-stop facility whilst also having replacement combs available that make through dovetails possible.

At the top end of the scale, the more expensive jigs start to give you every type of control regarding the angle of pitch, and board width and thickness. To all appearances, fine joints cut on these jigs are indistinguishable from good hand-cut joints. This means that for design-led makers and professionals, the price is justified in terms of the quality of their work. The Leigh jig utilizes a set of fully adjustable fingers that have infinite possibilities, whereas the likes of the Trend DC400 offer a range of templates and give more limited opportunities for variation.

 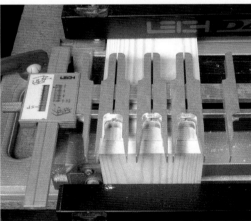

ABOVE When using the Leigh jig, each part of the joint is worked separately – first machine the tails, then flip over the template and machine the pins. This is a more complex jig to use, but the adjustable 'fingers' also makes it extremely versatile.

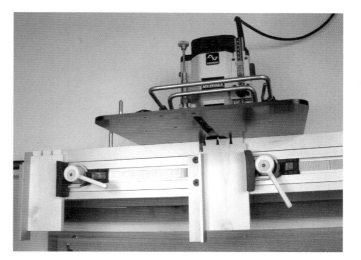

ABOVE The WoodRat is a versatile jig that can be used for a number of tasks, and is unique in that the workpiece is routed in the same direction as the cutter rotation. While this shot shows the tail piece of a through joint, the WoodRat can also be used to work mortise and tenon joints.

sizes of joint possible for stock of 75–125mm and for angles from 0–90°, the jig also offers the facility for up to quadruple mortise and tenons. Despite its scope for variation, however, the Leigh FMT is only a little more difficult to use than the simple but excellent Trend jig.

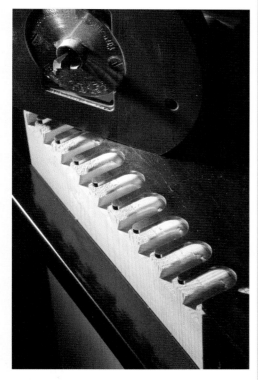

ABOVE With both workpieces clamped in place, both parts of the joint can be routed at once. Although this fixed-comb jig is used with a single dovetail cutter, some jigs require the use of a dovetail and a straight cutter to produce a joint.

KEY POINT

If you are considering buying a jig, I would strongly advise you to undertake your own research to ensure that you get what you need. As with all woodworking kit, the best way to decide is by watching demonstrations at a woodworking show.

KEY POINT

2:5 Joints

A wide range of joints can be made with a router, the best known of course being the dovetail. However, since dovetail jigs are covered in great detail elsewhere (see chapter 2:4, page 121), this section is devoted to what will ultimately be more useful and versatile constructional joints.

The six most commonly used router joints are the lap, housing, mortise-and-tenon, box comb (finger), biscuit and profile-and-scribe joints, most of which can be made using the jigs described in part three of the book (see page 149). There is virtually no design that cannot be achieved using versions of these joints and so I would strongly advise making the jigs first. This will ensure that time pressures do not force you into going for the quickest, easiest construction solution.

There are of course methods of jointing that can be achieved with a router, such as tongue-and-groove joints, loose tongues, lock mitres and spline joints, but the following joints will more than suffice as an introduction to routing.

Lap joints

Cutting lap joints with a router requires no jig, but is best undertaken on a router table. The lack of a fine height adjuster will make the task very difficult, so if you haven't got this vital accessory, buy or make one first. The only prerequisites are that you use workpieces of the same thickness and that your push board is cut at exactly 90°. Make sure that you perfect the set-up by practising on some off-cuts that are the same dimension as the timber you intend to use.

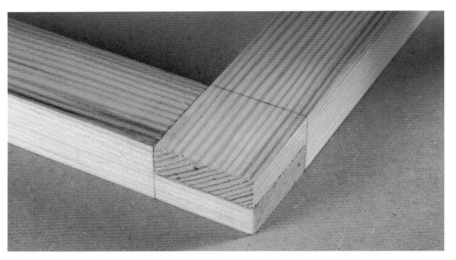

ABOVE A 90° lap or halving joint.

ABOVE Making the first pass.

Preparation and method

- Cut a pushboard with a corner of exactly 90°. The board should be thicker than half the thickness of your workpiece. This is in order to support the cutter's exit and thus prevent breakout.
- Set the fence so that its position acts as a stop to produce the joint's shoulder.
- Fit a straight cutter (either 12mm or 18mm is best) and adjust its height so that it removes an appropriate amount with each pass.
- Machine the workpiece in stages at this height setting before adjusting the height, again in stages, until it reaches the halfway line. The resulting joint should be absolutely flush.

ABOVE The lead-in fence acts as a length stop for routing the joint's shoulder in the final pass.

ABOVE The finished components.

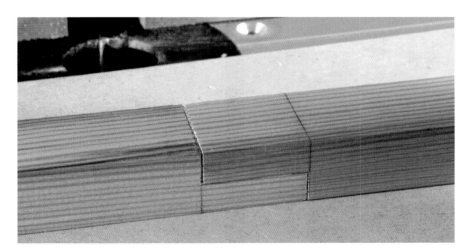

ABOVE The lap can also be used as a basic end-to-end joint but will have limited mechanical strength other than given by screws and or glue.

Housing joint

This joint is made using the jig featured on page 157. Through, stopped and dovetail housings can be accurately cut using this simple jig and, with addition of an appropriate spacing bar, consistent repeat cuts are made easy. The ideal housing set-up is that you'll have prepared your timber to be 0.2–0.5mm thinner than your chosen cutter's diameter. This will ensure that you get a perfect fit by only making one pass. However, those of you without a thicknesser will need to use some sort of shim or spacer (after the first pass) between the router base and the fence so that a second pass cuts the final dimension. Of course, if you use an enclosed guidebush jig (as shown below right), the width of the slot and diameters of the cutter and guidebush will all combine to produce a foolproof cut.

It is easy to achieve good tolerances simply by eyeing the position of the jig, but for equidistant housings, fix a spacing bar to the underside of the jig. This will locate easily into your first housing and give the perfect distance to the next cut.

The jig shown was made specifically for the cutter and router being used. Whether you are routing through, stopped or dovetail housings with a version of this jig, the principles are the same.

ABOVE Screwing a spacing bar (of the same width as the housing) to the underside of the jig.

ABOVE Using the housing jig described on page 157. The router's base simply runs against the jig's fence. For stopped housings simply add a stop block of some sort.

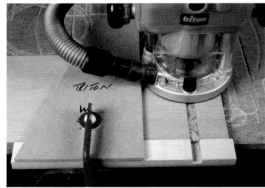

ABOVE Simple housing jigs such as this are designed to work with specific routers and cutters.

Preparation and method

- Mark out your timber.
- Cramp the jig to the workpiece and the bench. Ensure the jig's batten fence is pushed firmly against the workpiece.
- Rout the housing by running the router's base along the jig's fence. For through housings, start at the front edge; for stopped housings you should rout from the back edge.

⒡FOCUS ON:

Dovetail Housings

Because dovetail cutters cannot be plunged in stages, the cut must be made in one pass. The end-tail board should be cut on a router table, whereas the housing socket is best cut using a housing jig with a guidebush slot. Cut the housing first and add paper shims to the surface of the table's fence in order to get a perfect sliding fit.

ABOVE A range of dovetail cutters and their housings.

ABOVE A well-cut dovetail housing requires no glue and will be able to cope with movement in the timber.

RIGHT The most failsafe way of routing a housing without making mistakes is to use a guidebush located within a slot. This slot can of course be wider than the guidebush in order to rout housings wider than the cutter.

ABOVE The tail part of a dovetail housing is best cut vertically on a table. Obviously the board's end must be cut dead square.

Mortise and tenon

TECHNIQUE:

Preparation and
Method

The best way to cut a mortise is with a mortising machine. However, if you don't have the funds, your router can make a good job of a mortise moderately quickly. For tenons, a table set-up is ideally suited.

Cutting a tenon on a table is easy once you have perfected your technique. A fine height adjuster is essential since the difference between a perfectly fitting tenon or not is about 0.1–0.2mm on each tenon cheek. Because of the stresses placed on the workpiece during cross-cut routing on a table, it is best not to rely on a mitre fence – these invariably have a certain amount of 'play' within their guide channels. Instead, I always rely on a square-ended push board that is run along the table's fence.

ABOVE A mortise and tenon – both produced with a router.

TECHNIQUE:

Preparation and Method

Note: the following preparation-and-method guidelines relate to the step-by-step instructions featured on pages 142–5. The mortise is made using the mortise jig, construction details for which can be found on page 165; the tenon is simply cut on a router table. It is also possible to rout a mortise and tenon joint using a commercial jig such as a WoodRat (see chapter 2:4, page 121, for more information).

Mortise

- Screw the mortise box to the bench before packing up the workpiece flush with the top edge and cramping it to the box side.

- Position (either by calculations or by offering up the router) and pin the length stops across the top of the box. Transfer the relevant workpiece markings to the mortise box, in order to rout any repeat cuts.

- Adjust the parallel fence to align the cutter.

- Set the depth to approximately 2mm more than the tenon length.

- Plunge and rout in a suitable number of stages.

Tenon

- Mark up a test workpiece and then, by trial and error, set the height of the cutter to cut a tenon.

- Position the fence at a distance from the cutter that equates to the length of your tenon. You may have to add a false face in order to achieve the correct distance.

- Select a flat, square-ended 'support board'. Cramp your workpiece so that its end is precisely flush with the edge of the support board. Make sure that the cramp is secure and well clear of the cutting area.

- Make your first shoulder-cutting pass (using a ½in cutter), then flip the workpiece over and rout the opposite side. Next, cramp the workpiece on its edges and rout the sides of the tenon.

- Once this whole 'rotated' pass has been completed, move the workpiece away from the fence in stages and rout the next rotation, and so on until it is complete.

- To match the rounded ends of the mortise you will need to shape the ends of the tenon. Although not difficult, care should be taken to keep this paring-off square, particularly with anything other than perfectly straight-grained stock. Because of its small teeth and set, I find that a dovetail saw or even a junior hacksaw is ideal for cutting the base of the waste without causing damage to the shoulder.

- The completed joint should require no more than a couple of taps with a mallet to seat properly. Any more and the joint is too tight.

Combing joint

In many ways, I prefer the detail of a comb joint to that of a dovetail, and they are especially pleasing when you have made the jig yourself. This joint is made using the jig featured on page 168, although combing joints can also be cut using an independent combing template or one compatible with a dovetail jig. Plywood as well as solid timber is appropriate for this type of jointing, the decorative qualities of which can be accentuated by mixing contrasting timbers. By replacing the locating pin with a lower, appropriately sized version it would be possible to use this jig for routing dentil mouldings or similar decorative features.

ABOVE The decorative qualities of a comb joint can be accentuated by mixing contrasting timbers.

ABOVE Butt the workpiece up against the guide pin and make the first pass.

The thickness of each comb segment or finger is up to you, but the norm is 6–12mm (requiring a cutter of an equal diameter). If you are designing a piece of furniture that uses finger joints, you must first decide on the thickness of the proposed fingers so that your drawers, frame constructions and so on have the relevant dimension divisible by this amount. In the example shown I have used a ¼in diameter cutter to rout comb joints in some perfectly square stock.

ABOVE Locate the first combing slot over the pin and rout the second pass.

ABOVE Continue the process.

Preparation and method

- Set the height of the cutter to be 1–2mm greater than the thickness of the stock. This will produce over-length fingers that can be trimmed back neatly using the simple flush-trimming jig featured on page 162 or a block plane.

- The fence of the jig will need to be adjusted so that the nearest face of the locating pin to the cutter is exactly ¼in away from the outer perimeter of the cutter. The easiest way to set this is by using another ¼in cutter.

- For the first cut, butt the workpiece up against the locating pin and push the fence assembly carefully over the cutter. Note: the end of the workpiece must be cut absolutely square if it is to be presented properly.

- The second and all subsequent cuts are simple – fit your last slot over the locating pin in order to rout the next slot and so on.

- Test the resulting fit of the set-up by routing some narrow stock (to make it quicker) and then test fit the joint. Minute adjustments in the fence make the difference between a loose joint and one that is too tight to fit together, so be patient in the set-up and you will be rewarded with easy to assemble, perfectly snug joints.

Profile scribing

There are two types of profile-and-scribe cutters: combination and all-in-one. Combination cutters must be reassembled to cut both parts of the joint, whereas all-in-one cutters are taller and so can cut both parts of the joint simply by being raised or lowered.

For the purposes of this book I will demonstrate the combination-cut joint, which despite sounding complicated is actually very easy to machine. This type of cutter always comes supplied with instructions and I would advise referring to these as well. For a full step-by-step guide, see page 146.

ABOVE Frame and panel constructions are made easy by using profile-and-scribe cutters, which cut the joint, profile and panel groove in two passes.

TECHNIQUE:

Preparation and Method

Note: the following preparation-and-method guidelines relate to the step-by-step instructions on pages 146–7.

- The first stage is to set up and rout the scribe joint. For this you'll need to construct a 6mm MDF sliding carriage with a crosscut fence. This will move the workpiece squarely past the cutter and prevent breakout from the back edge. In dimension, the MDF carriage should be up to half the length of the table and as deep as the table top in front of the fence.

- Attach a perfectly square batten (as thick as your workpiece) to the MDF at exactly 90° to the edge.

- Line up and square the table's fence with the cutter's bearing guide and then raise the cutter so that its bottom cutting edge will remove a 1–2mm rebate from the surface of the MDF when it is passed along the fence. This machining will also cut the profile through the end of the batten fence.

- Cramp your workpiece to the carriage fence so that it butts up against the table's lead-in fence and then make a pass through the cutter. If satisfactory (check the manufacturer's instructions), scribe-cut all of the other cross rails for the job at the same time.

- Switch off the power, disassemble the cutter (as per the manufacturer's instructions) and adjust its height using the flipped-over scribe-cut piece as a guide.

- Fit your table's hold-downs and test cut a profile. You may need to make some very fine adjustments to the height of your cutter in order to achieve the perfect fit.

- Once you have the correct height setting you can profile the stiles and of course the cross rails that you scribed earlier.

- Once complete, the assembled joint should be a perfect fit.

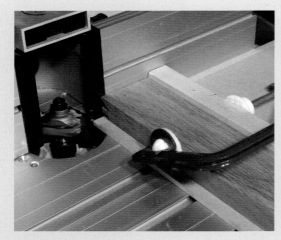

ABOVE The set-up used – a 6mm MDF sliding carriage with a perfectly square crosscut fence.

ABOVE By using this set-up, the workpiece is moved squarely past the cutter and breakout is prevented at the back edge.

Biscuit joints

If you don't have a biscuit jointer but want to utilize this excellent jointing system, you can always use your router and a 4mm slotting cutter. I always find it easier and safer to rout edge slots on a table, but freehand slotting is perfectly acceptable.

Although horizontal-slotting cutters can't rout board faces as biscuit jointers can, by using a 4mm straight cutter (in a router, run against a fence) this problem can be overcome. However, if you are intending to undertake a fair amount of carcass work, you should seriously consider the quicker and easier option of buying a biscuit jointer.

Edge jointing

Arrange your components precisely in position and then mark the positions of your biscuits by simply drawing a line

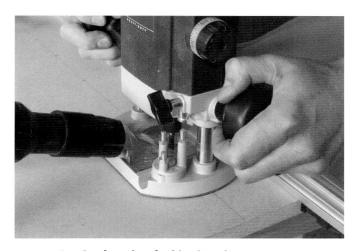

ABOVE Routing face slots for biscuits using a 4mm straight cutter.

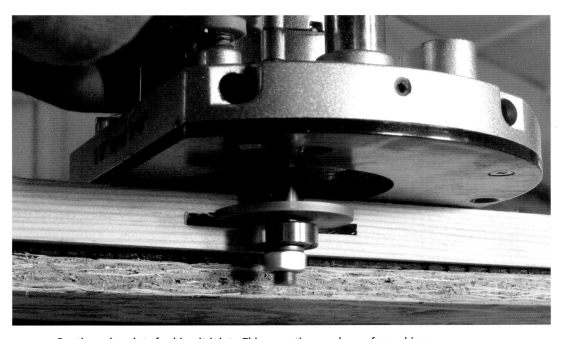

ABOVE Routing edge slots for biscuit joints. This operation can be preformed in a table or freehand (as shown here).

ABOVE Marking up the biscuit centres.

across the meeting faces. Because the diameter of biscuit-slotting cutters is normally 40mm, in order to rout suitably

long slots for your size of biscuit you will need to move the router from left to right. Start from left of the centre mark and ensure that the router stays at the plunged depth. The amount you'll need to increase the slot's length by depends on whether you are using '0', '10' or '20' biscuits. Don't worry if you make the slots too long – they can be over-length, but, as described below, there is a functional minimum. Test this technique to make sure and, if necessary, add some extra marks on either side of the centre marks to denote the amount of sideways movement needed.

When table routing you will need to push the workpiece onto the cutter and then move it to the left once it is against the fence, following the instructions for the amount of side-to-side movement

FOCUS ON:

Slot Sizes

FOCUS ON:

Slot Sizes

The following slot sizes are over-sized by approximately 12mm and will give you all the room you need for each biscuit:

'0' biscuits: 60mm long slots
'10' biscuits: 65mm long slots
'20' biscuits: 70mm long slots

RIGHT An oversized (i.e. 70mm) slot has been routed for this '20' biscuit.

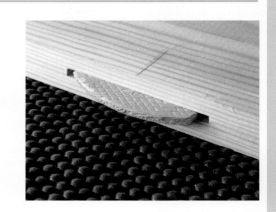

KEY POINT

KEY POINT

As many slotting cutters have a short shank, it may be necessary to use a collet extension in a table. This will raise the slotting teeth to a sufficient height, even for 19mm stock.

relates to the movement needed. However, as I have said, don't worry too much about making the slots too long – too long is better than too short.

needed (see panel). If you do not grasp the workpiece firmly there is a risk that the cutter will grab it and throw it to the right. However, if you stand firmly in front of the table and your movements are decisive, you will encounter no problems.

If you have trouble getting the slots long enough, mark the cutter centre on the fence and then add a mark either side that

ABOVE Once pushed onto the cutter, the workpiece will need to be moved from side to side in order to lengthen the slot.

TECHNIQUE:

Carcass Assembly

TECHNIQUE:

Carcass Assembly

In order to rout a combination of face and edge slots, you will need to use both horizontal slotting and straight (4mm diameter) cutters. The same rules regarding slot length will apply to the edge and face. The hardest part of this kind of combination routing is positioning the biscuits correctly in relation to each other, but with practice you will develop an accurate system.

ABOVE Presenting the workpiece centrally to the cutter on a table.

Making a mortise-and-tenon joint

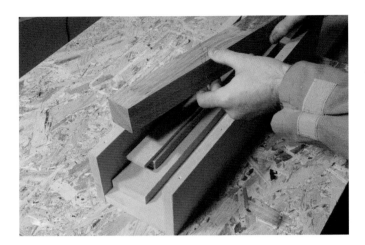

1 Pack up the workpiece so that it is flush with the top of the jig. (For more details on the preparation and methods used for this step-by-step guide, see pages 134–5.)

2 After cramping the workpiece to the side, position and pin length stops at 90° across the top of the box.

3 Rout the mortise to depth in an appropriate number of passes.

4 The completed mortise.

5 The set-up for cutting tenons on a router table. The workpiece is cramped to a perfectly square support board. This assembly is run against the lead-in fence (face board added to achieve correct tenon length), with the workpiece being re-clamped for each face.

Making a mortise-and-tenon joint

6 Making the first shoulder pass.

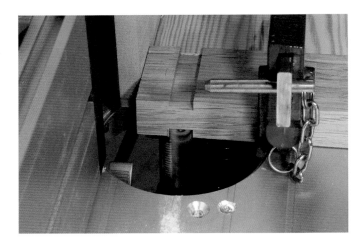

7 Making the second pass.

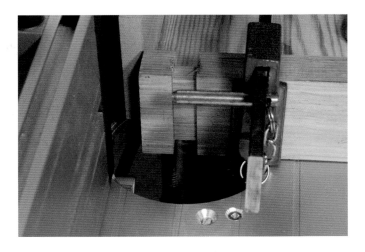

8 Cutting the final shoulder.

9 Beginning the rest of the passes.

10 The machining finished.

11 The corners of the tenons will need to be rounded off. First cut the curve at the shoulder end of the tenon using a fine-toothed saw, before paring down with a chisel.

Making a profile-and-scribe joint

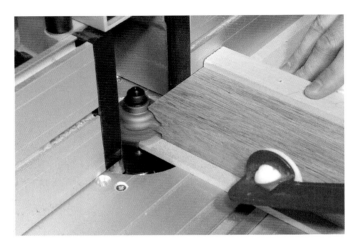

1 Routing the scribe using the set-up shown earlier. (For more details on the preparation and methods used, see page 138.)

3 Flip over the workpiece and adjust the cutter's height so that the groover lines up exactly with the tongue.

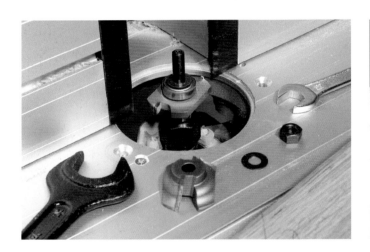

2 Disassemble, reconfigure and reassemble the cutter (following the manufacturer's instructions) to cut the profile.

4 Fit your hold-downs and then make some test cuts to check that the cutter's height setting is correct.

5 Routing the profile.

6 Profiling a previously scribe-cut cross rail.

7 The completed parts.

Part 3:
Projects

Large router table

Despite costing very little to make, this router table will work as well as a proprietary set-up – providing you spend the time getting every detail right.

In my experience, it will take the average beginner up to four days to make this project. The small circle jig (see page 158) is needed during the construction process.

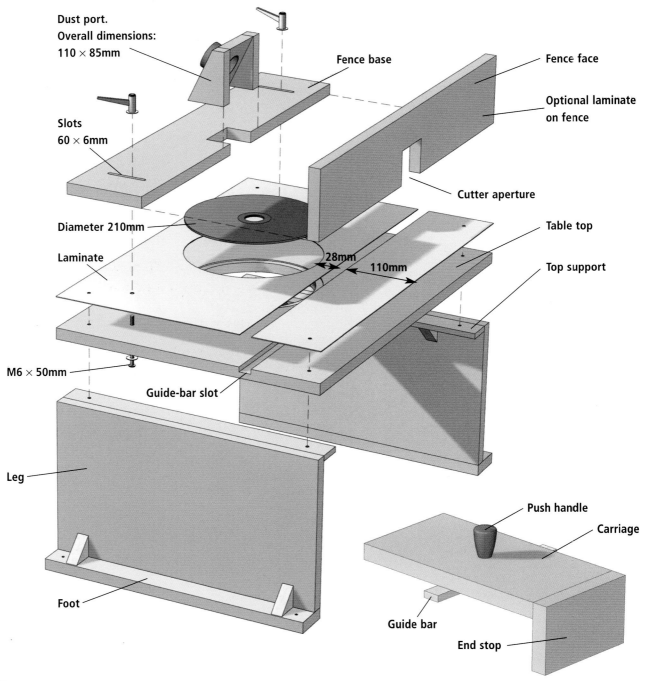

Dust port.
Overall dimensions:
110 × 85mm

Fence base

Fence face

Optional laminate on fence

Slots
60 × 6mm

Cutter aperture

Diameter 210mm

Table top

Laminate

28mm

110mm

Top support

M6 × 50mm

Guide-bar slot

Leg

Push handle

Carriage

Foot

Guide bar

End stop

Cutting list

Table:

• Top	600 × 430 × 18mm
• Top supports (×2)	430 × 40 × 18mm
• Legs (×2)	430 × 324 × 18mm
• Feet (×2)	430 × 60 × 18mm

Fence:

• Face	600 × 110 × 18mm
• Base	600 × 140 × 18mm

Sliding carriage:

• Carriage	230 × 120 × 18mm
• End stop	120 × 75 × 18mm
• Guide bar (hardwood)	150 × 20 × 8mm

Construction

Components

Cut the component parts slightly oversized and then true the edges with a hand plane. Make sure that they are square and square-edged.

Pre-lamination

Fix the two parts of the fence together, making sure that they are straight and that they rise at exactly 90° from the table.

Use contact adhesive to stick the laminate to the fence and the table top, being careful to not use too much glue.

When properly stuck, use a bearing-guided trimmer to clean up the edges and then add a chamfer to the table top's laminate's edges.

Insert plate

The advantage of using a thin metal insert plate is that it maximizes the potential height of the cutter. Secure the table top to a sacrificial board, making sure that the centre of the circular cutout is stuck with double-sided tape.

Using your small circle guide and a ½in diameter cutter, rout a recess as deep as the insert plate is thick (+0.5mm). The diameter of this circle should be 1–2mm greater than that of the insert plate.

Re-set the small circle guide to cut a diameter of 20mm less than that of the insert plate and fit a ¼in diameter cutter.

Plunge cut through the table top in stages of anything up to ¼in (6mm). Once finished,

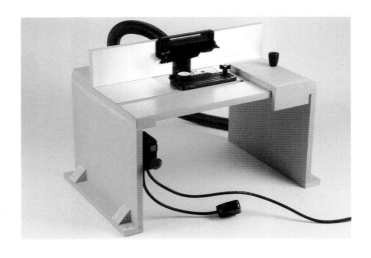

Large router table

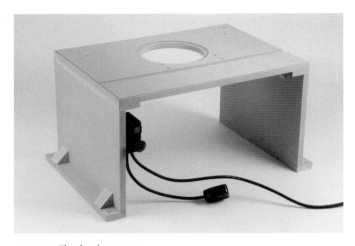

ABOVE The basic carcass.

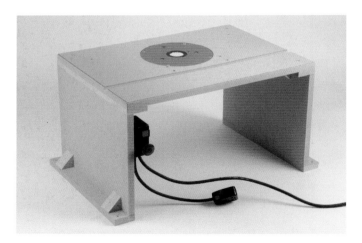

ABOVE The carcass with the insert plate added.

remove the cut-out and fix the insert plate into its recess.

Sliding fence groove

Rout the groove for a cross-cut carriage using a parallel fence and a cutter of a suitable diameter so that it can be routed in one depth pass. If you do make a mistake, you can always widen the groove in another pass and adjust the dimensions of the inserting rail to suit.

Fence

First, you'll need to rout out the cutter aperture using a template and a guidebush. Once you've routed one face, carefully align the template on the other face and make a repeat cut.

Next, rout the slots for the fence fixing and adjustment bolts. A ¼in diameter cutter will provide a perfect loose fit for an M6 thread.

The extraction-port construction can simply be butt-jointed with a little glue, but before you start you'll need to cut the hose aperture in an oversize workpiece using the small circle jig. Once you have achieved a good fit, trim this workpiece down to size. Then attach the side panels and cut the construction to shape.

Once dry, the extraction port assembly can be butt-jointed and glued into position.

Legs

The leg frames can be assembled separately and then attached by screwing through the table top.

45° angle fillet blocks must be added in order for the table to cope with the stresses of machining. These can again be rub-jointed with glue.

The whole construction is too high for use on a standard workbench, so you will need to find a suitable sturdy support table as a temporary or permanent home.

Sliding carriage

Rather than make a multi-angle parallel fence, I thought it more useful to make a rock-solid 90° fence that could be used to cut the likes of halving joints. Often, however, a simple support board run against the fence will be the best solution.

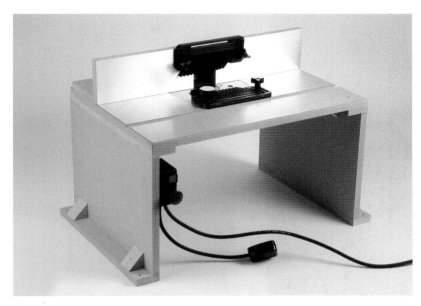

ABOVE The main fence added.

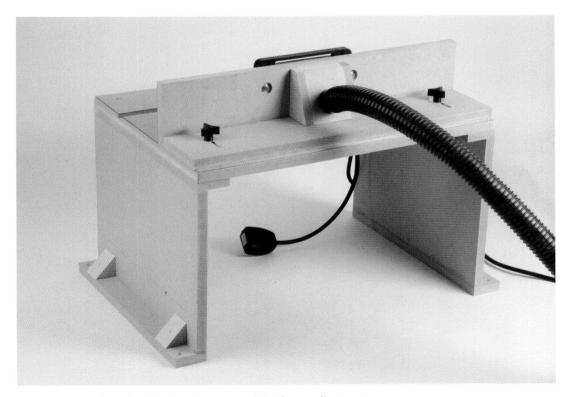

ABOVE A rear view showing the dust port and the fence adjustment.

Small router table

This project should demonstrate just how simple a router table can be to make, and although this bench-top version is only really suitable for small-dimension timber, the same design could be scaled up to the same dimensions as the large table (see page 150). As previously, the small circle jig (see page 158) is needed during the construction process.

Cutting list

Table:
- Top (×2)   300 × 200 × 6mm
- Legs (×2)   250 × 200 × 18mm
- Base   400 × 200 × 18mm

Fence:
- Face   300 × 70 × 18mm
- Base   300 × 80 × 18mm

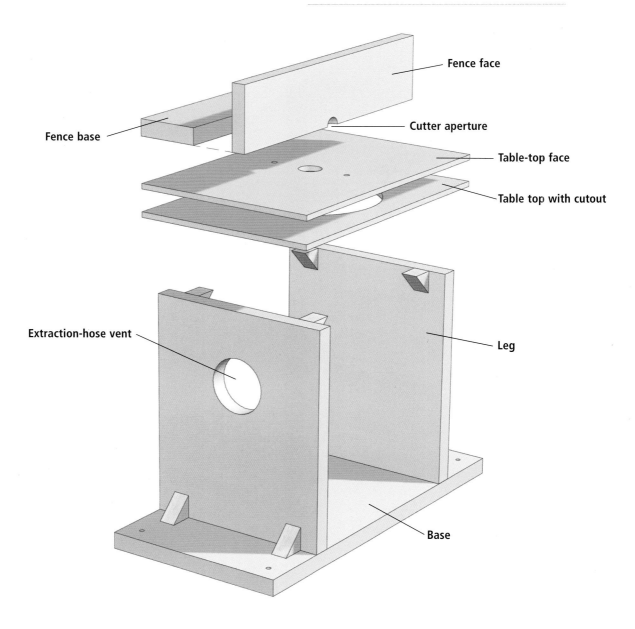

Fence face

Fence base

Cutter aperture

Table-top face

Table top with cutout

Extraction-hose vent

Leg

Base

Construction

Components

Cut the component parts slightly oversized and then true the edges with a hand plane. Make sure that they are square and square-edged.

Table top

The table top is constructed of two pieces of 6mm MDF that are glued together after a circular cutout is machined from the lower piece.

Using the small circle guide, rout a piercing circular cut through a workpiece that is stuck to a cutting table with double-sided tape. The diameter of this circle should be around ¼in (5mm) greater than the dimension of your router's base.

Glue the two parts together and, after the assembly has dried, stick the router centrally inside the recess using double-sided tape. Using a straight cutter, plunge a hole through the table top and then remove the router.

With the cutter protruding slightly through the base, locate the cutter in the hole, but this time on the top of the table. Square the router to the side of the MDF and draw around its base. Transfer the position of the base-fixing screws onto the table top before drilling and countersinking.

Legs

Using the small circle template, cut out an oversized hole in the appropriate leg for a dust-extraction hose.

Assemble the frame and construct the fence, making sure it is absolutely square.

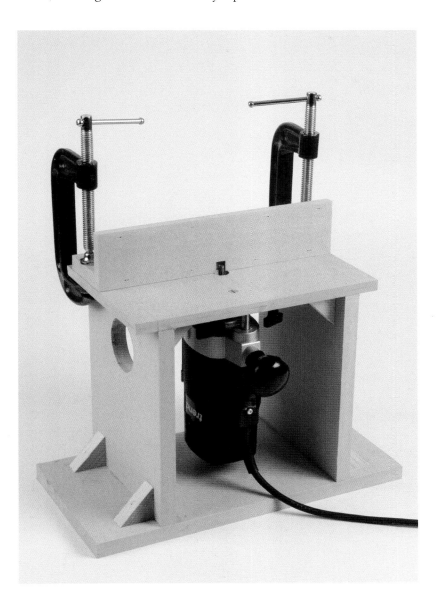

Small router table

Once the glue is dry, cramp the fence to the table top with its face directly over the centre line of the cutter, before plunging the previously used straight cutter up into the fence.

Whatever diameter cutter is used in future, the hole in the fence and through the table top will need to be approximately 25% bigger or more in order to aid extraction. Without being present in the fence, this will need all the help it can get to extract through the dust spout.

ABOVE The basic carcass.

ABOVE A rear view.

Housing jig

The use of this jig is demonstrated on page 132. Through, stopped and dovetail housings can be cut accurately using this simple jig. Also, with addition of an appropriate spacing bar, consistent repeat cuts are made easy. The jig is, of course, set up for use with a particular router and cutter combination. However, because it is such a simple construction, there is no reason why you can't make two or three at one sitting.

Cutting list

- 400 × 350 × 18mm MDF
- 550 × 30 × 20mm softwood

Construction

Once you have glued the fence to the main board, use the jig to make a pass into some scrap wood. This will create the channel through the fence batten.

The cutter can then be started within this channel for future cuts, giving extra support for the router.

ABOVE A length stop can be added for equidistant housings.

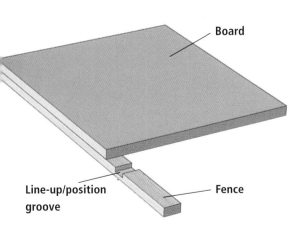

Board

Line-up/position groove

Fence

Small circle jig

This wonderfully simple jig can cut perfect circles down to about 30mm diameter. The results are easy to control, and the set-up takes only a few seconds. I can guarantee that if you don't make this jig right away, you'll be kicking yourself later.

Cutting list

- 255 × 167 × 6mm

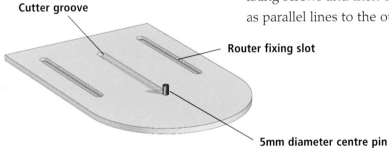

Cutter groove

Router fixing slot

5mm diameter centre pin

Construction

Cut out a piece of 9mm MDF to whatever length you require (and slightly wider than the widest part of your router's base).

Mark a centre line down the length of the jig and position the router so that its spindle is exactly centred on this line. Mark the position of your router's base-fixing screws and then extend these points as parallel lines to the other end of the jig.

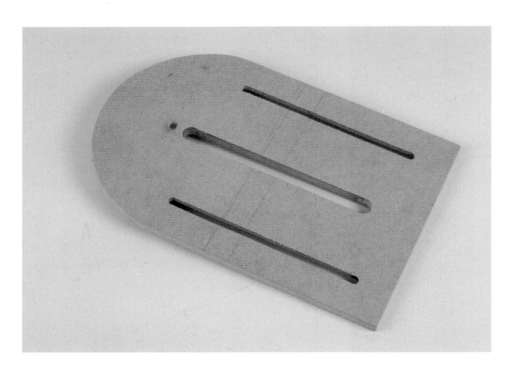

Attach a parallel fence to your router and fit a 6mm or ¼in cutter. Rout a piercing slot centred on the closest fixing-screw lines to within about 30mm from the far end. Without adjusting the fence, fit a 12mm or ½in cutter and rout the same centre line to a depth of 5mm, then extend the length of cut by a further 3mm at each end. Repeat the process for the opposite side.

With the 12mm or ½in cutter still fitted, rout a piercing slot down the jig's centre to 30mm from the spindle centre point.

Cut a short length of steel or aluminium and place it in a pre-drilled hole about 20mm from the end of the central slot. Glue it in place so that 5mm stands proud of the jig's underside.

The jig can now be attached to your router using pan-headed screws whose heads recess into the external slots.

I find that by using a pair of Vernier callipers to set the distance, the results are normally perfect first time.

ABOVE Attached to the router.

ABOVE For routing arcs and full circles.

End-trimming jig

This is probably the simplest jig of all, but it produces perfectly trimmed end grain. Remember to rout from left to right in order to prevent the cutter grabbing the workpiece and giving unwanted breakout. Care must also be taken that the jig's fence is tight up against the workpiece and that the cramp is exerting enough pressure to prevent movement.

Cutting list

- Board   160 × 160 × 9mm
- Router fence   160 × 30 × 20mm
- Base fence   160 × 30 × 20mm

Construction

Cut the board slightly oversized and then true two edges perfectly square with a hand plane. Make sure that the two battens are square before carefully attaching them to the board.

The workpiece fence must be absolutely square to the router fence and the only way to check this properly is to make some test cuts.

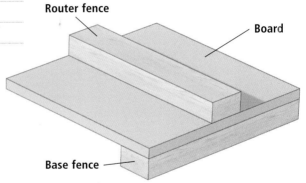

Router fence

Board

Base fence

ABOVE The jig is cramped onto the workpiece and the bench.

ABOVE Trimming the end flush.

Flush-trimming jig

This perfectly simple jig makes short work of trimming the likes of solid lipping, through-tenons, dovetails and comb joints to within about 0.1mm of flush. The best way to set up the jig is to plunge the cutter (with the motor off) onto a piece of paper or card and then lock the depth stop. The subsequent cut will be the thickness of the paper proud of the surface.

However, the jig isn't foolproof – it is possible to tip up, letting the cutter bite into the workpiece. Care is therefore needed. Alternatively, if the job allows, use double-sided tape to add a support block to the underside of the router's baseplate. Also, to reduce the risk of damage whilst

machining, be conservative with the amount you take off with each pass.

Cutting list

- Base   250 × 160 × 9mm
- Sub-base   160 × 160 × 9mm

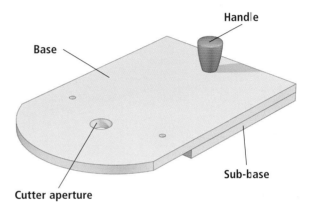

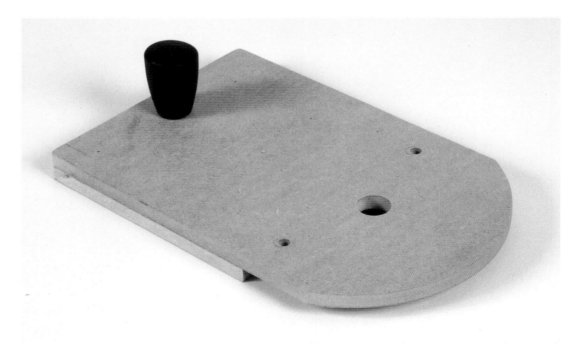

Construction

Mark out the sub-base outline, fixing holes and cutter aperture.

Cut out the sub-base and drill the holes.

Cut out a base plate that is just short enough to allow the countersinks for the router fixing screws to be clear of its edge.

Glue and cramp the sub-base and base plate together.

Fix a handle to the back end of the jig.

If you are using a ¼in router, a ½in or ¾in diameter cutter is ideal.

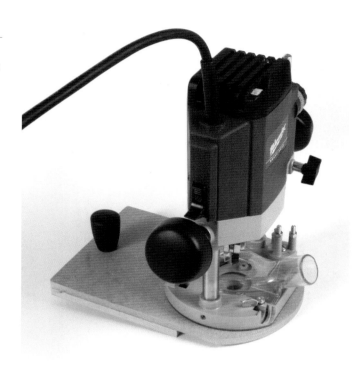

ABOVE The router attached.

ABOVE The sub-base can extend as far as the perimeter of the cutter.

Flush-trimming jig

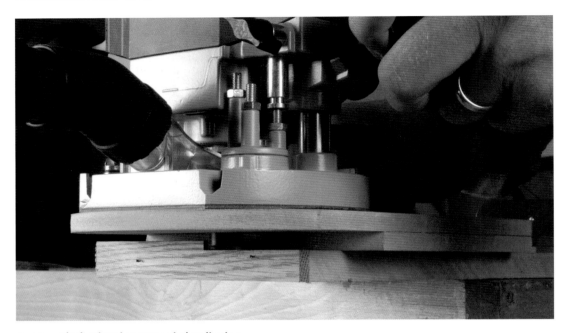

ABOVE Flush trimming some timber lipping.

LEFT The result: a perfectly flush cut.

Mortise jig

The use of this jig is demonstrated on page 142. As you can see, it couldn't be simpler. The box can be any length or width, but I found the dimensions shown to be the most useful. The tenon is best cut on a router table, as explained in the chapter on joints (see page 134), so this project deals specifically with the mortise.

Cutting list

- Base   500 × 100 × 18mm
- Sides (×2)   500 × 100 × 18mm

Construction

Cut the component parts slightly oversized and then true the edges with a hand plane. Make sure that they are square and square-edged.

Assemble the sides and the base and then cut some suitably sized battens as stops.

It is a good idea to store useful packing with the jig so that achieving a flush workpiece doesn't become difficult.

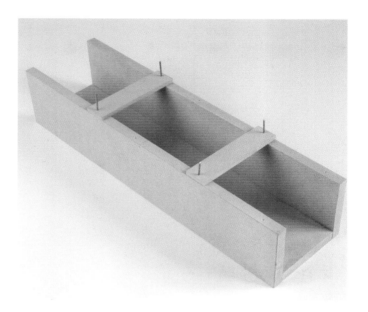

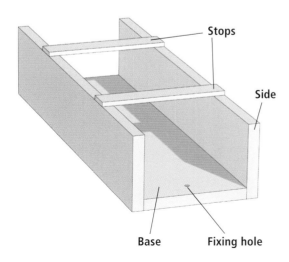

Stops

Side

Base   Fixing hole

ABOVE In use – note the length stops.

Trammel jig

This basic circle and arc-cutting jig is quick to make and will suffice for the majority of jobs. I have pre-drilled the distance holes at 5mm increments in this instance, although there is nothing to stop you drilling the holes as and when you need them. Three holes are needed at the router end – one as a cutter aperture and the others for the machine screws that fix the jig to the router's base. If you want to make the jig fit a different-sized router, simply rotate the new router's position 90° and drill the alternative fixing holes. The most straightforward method of locating the centre of the jig is to use the same drill bit as used to drill the centre point of the workpiece.

Cutting list

- 750 × 120 × 6mm MDF

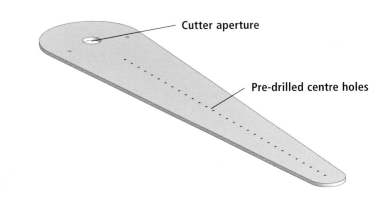

Cutter aperture

Pre-drilled centre holes

ABOVE The set-up.

ABOVE A drill bit can be used as the centre locator.

ABOVE Routing an arc.

Box-combing jig

This is a relatively complex jig for such a simple joint, but since it is an adjustable set-up, you'll be unlikely to ever need another medium-sized jig. The jig featured is set up to rout ¼in segments but could take up to a ½in diameter cutter if a new ½in guide pin was fitted to a replacement face fencing. The use of a surface laminate is optional, but on the table top at least, it does aid the smooth movement of the fence. MDF worksurfaces are perfectly adequate. The small circle jig (see page 158) is needed during construction.

Cutting list

Table:

- Top   350 × 300 × 18mm
- Legs (×2)   300 × 230 × 18mm
- Feet (×2)   300 × 70 × 18mm

Fence:

- Base   300 × 100 × 18mm
- Front   300 × 100 × 18mm
- Facing   300 × 125 × 18mm
- Guide bar   180 × 20 × 7mm hardwood
- Indexing pin 25 × 25 × 6mm hardwood

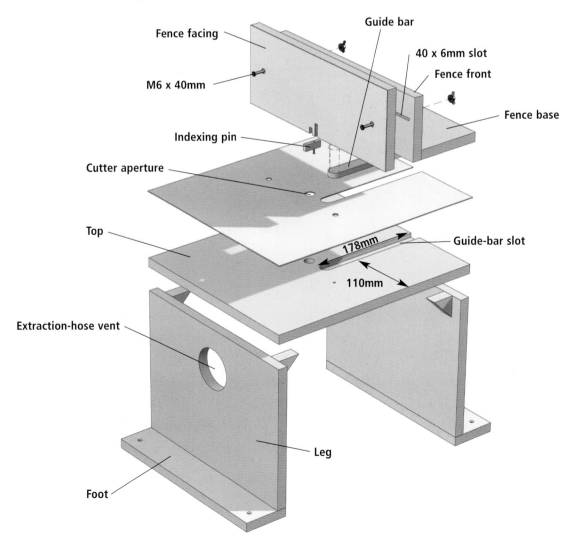

Fence facing · Guide bar · 40 x 6mm slot · Fence front · M6 x 40mm · Fence base · Indexing pin · Cutter aperture · Top · 178mm · Guide-bar slot · 110mm · Extraction-hose vent · Leg · Foot

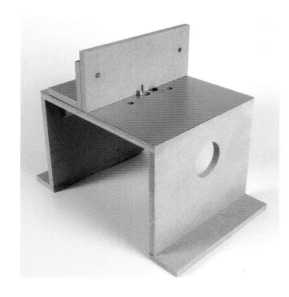

Construction

Components

Cut the parts slightly oversized and then true the edges with a hand plane. Make sure they are square and square-edged.

Lamination

Use contact adhesive to stick the laminate to your chosen parts, being careful to not use too much glue. When properly stuck, use a bearing-guided trimmer to clean up the edges and then add a chamfer to the laminate's edges.

Router position

Measure and drill the holes needed for the router-fixing screws and then countersink the holes with care. Fix the router in position and then, using a ½in diameter cutter, plunge through the table top.

Fence and guide pin

The fence assembly must sit at 90° to the table top.

Join the back fence and its base plate, being careful to avoid fixings around where the cutter will run through.

Rout the slots in the back fence assembly and drill the fixing holes in the front fence.

Guide bar and slot

Rout the slot in the table top using a ¾in diameter cutter in a single pass, then thickness a guide bar to fit using the methods described for the guide pin. Approach the final width dimension very carefully as you will need a perfect tight fit. Screw the guide bar to the fence base plate at exactly 90°.

ABOVE **The main carcass and fence assembly.**

Box-combing jig

ABOVE The underside and back of the fence.

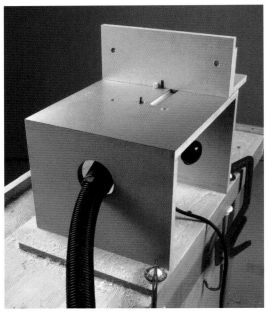

ABOVE Assembled and ready for action.

If made correctly, the fence assembly will only move in the desired direction. There should be no twisting within the slot.

Adjust the position of the sliding fence and rout the slot for the guide pin, moving the cutter all the way through the sliding fence. This slot and its pin should only be as high as the thinnest material you intend to join with the jig minus a millimetre (5mm is probably best).

Make sure that the pin is exactly the thickness of your chosen cutter. If you don't have a thicknesser to dimension the stock for this pin you will have to use a plane and then resort to sandpaper for the final tweaks.

Legs
Use the small circle jig to rout a hole for a dust-extraction hose in the appropriate leg and then assemble the components. The angle blocks can be attached by rub jointing with some glue.

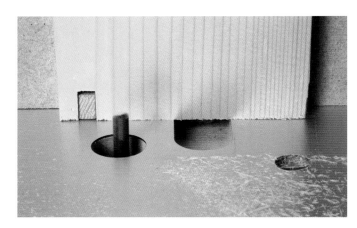

ABOVE After the first cut, the process is simple.

Glossary

B

Bearing-guided Cutter

A cutter with a rotating bearing guide mounted either above or below the blades. The cutter is guided by the bearing as the latter runs along the edge of a workpiece or template.

Breakout

Also known as 'spelching' or 'tearout'. Torn wood fibres created as a workpiece is routed. Wood can be particularly prone to breakout when routing across the grain.

Burning

Also known as 'scorching'. Dark patches on a routed workpiece caused by excessive friction. May be the result of a slow and/or inconsistent feed rate, or a blunt cutter blade.

C

Collet

A precision-made, high-quality metal sleeve comprising two or more segments, which are used to grip the cutter shank. The collet itself is inserted into the collet nut, linking the whole unit directly to the motor spindle of the router. A small but vital component in any router.

D

Depth Stop

The different stops on the turret onto which a plunge bar plunges to set the depth of cut.

Depth Turret

A revolving casting on the base of plunge routers. The turret features adjustable points that make it possible to alter the plunge depth in stages (most routers have a three-stage turret). Can be used in conjunction with a fine height adjuster for greater precision.

Dovetail

A strong, decorative joint formed by interlocking sets of pins and tails. Made in two styles: 'through' and 'half blind'. The latter is distinguished by the fact that the ends of the dovetail are concealed on one side.

Dust Cowl

A plastic port fitted to a router base to aid dust extraction. Connects to the hose of an extractor.

E

Elu

One-time king of router manufacturing, now subsumed by Black & Decker and DeWalt.

F

Feed Direction

The direction in which the router is passed through the workpiece. As a rule, the direction of cut should be against the rotation of the cutter.

Feed Rate

The speed with which the router is passed through the workpiece. The optimium feed rate should be fast enough to prevent burning but slow enough to avoid straining the motor and/or cutter.

G

Guidebush

A simple metal collar that can be fitted into a router base. Used in combination with an edge (such as that created by a template), this can be used to guide the cutter at a set distance from the workpiece.

H

Handheld Routing

Routing operations in which the router is used upright as a hand-held, portable power tool. The workpiece is held in a fixed position, while the router – and therefore the cutter – receives directional guidance.

High Pressure Low Volume

A purpose-built industrial vacuum extractor used for dust and chip removal when routing. Provides a higher level of extraction than High Volume Low Pressure models, which are suitable for static machinery such as table saws.

High-speed Steel (HSS)

A form of hardened steel used to make some router cutters. The term 'high speed' is derived from the fact that it is capable of cutting metal at a much higher rate than carbon tool steel.

I

Inverted Routing

Also known as 'table routing'. Routing operations in which the router is mounted upside-down in a router table. The workpiece is presented to the cutter, which is held in a fixed position.

M

Mortise and Tenon

A joint formed by a projecting end (tenon) that fits into a cut recess (mortise) of matching dimensions.

O

Offset

The difference between guidebush and cutter diameter. When using a template and guidebush in combination, the cutter follows the path of the template at an offset distance, which must be calculated in advance.

P

Parallel Fence

One of the most basic pieces of equipment used to control a hand-held router. The fence can be moved along a straight edge; by attaching it to the router base at a set distance using two adjustable rods, it will guide the cutter on a parallel path.

Plunge Depth

The depth to which a cutter is plunged. On a plunge-base router, this can be adjusted using a combination of the depth turret and the plunge lock. Workpieces are generally machined in several stages, using passes of increasing depth, to reduce stress on the motor and cutter.

R

Router

The ultimate power tool. Suited to most woodworking applications.

Rub Jointing

A means of edge-jointing timber using glue. With hand pressure, a workpiece is moved a small amout backwards and forwards for upwards of ten seconds. The joint will tighten once the air between the pieces has been expelled.

S

Sacrificial Board

Also known as a slave board. A cutting/routing table that can be cut into whilst machining a workpiece.

Shank

The precision-machined cylindrical section of a cutter that slides into the collet. The most common shank sizes (measured in diameter) are ¼in and ½in, but whatever the size this must be matched to a collet of the same width.

Stopped Routing

Also known as 'drop-on routing'. The practice of lowering a workpiece directly onto a rotating cutter in a table-mounted router. This technique requires good control of the workpiece and is not recommended for beginners.

T

Template

A routing pattern around or within which a cutter routs. Generally workshop-made from wood or plastic, a template provides a quick and easy way to duplicate any given shape with great accuracy and consistency. Templates can be used for both handheld and inverted applications, and are used in conjunction with a bearing-guided cutter or a guidebush.

Thicknesser

A planer thicknesser mounted in a woodworking machine. Used for squaring dimensions of timber.

Trammel

A simple point and arm device used to guide a router when cutting circles or arcs. Ellipses can also be formed using a trammel, although the arm must be linked to two points, not one.

Tungsten-carbide Tipped (TCT)

The most common cutter type. Only smaller TCT cutters are made entirely from tungsten carbide; normally, grains of the material are sintered (i.e. compressed under high temperatures) to form just the blades.

W

Workpiece

A piece of wood (or other material) that is in the process of being worked or made.

(FOCUS ON:

Conversions

The dimensions of the projects and jigs shown in this book are given in millimetres. These metric dimensions can be converted to imperial, should you prefer, using the table below as a guide. However, great care should be taken that your approximations are consistent and you should use either imperial or metric measurements consistently.

This also applies to manmade boards, which are often (in the UK) sold as metric but are in fact their imperial equivalent – i.e. '18mm' MDF is frequently more like ¾in (19mm). However, never make any assumptions about the thickness of a board and always measure it accurately before starting a project. With regard to cutters and shank and collet sizes, it is important to remember that their diameter is either metric or imperial – never both.

inches	mm	inches	mm
⅛	3	2	51
¼	6	2½	64
⅜	10	3	76
½	13	4	102
⅝	16	6	152
¾	19	12	305
⅞	22	24	610
1	25		
1¼	32		
1½	38		
1¾	44		

Index

About the Author

After qualifying as a carpenter in 1989, Stuart set up a joinery workshop in Brighton, UK, and through his work making furniture, developed an interest in design. This fascination eventually led him to study at the Royal College of Arts in London, UK, and to his current roles as a furniture designer and editor of GMC's *The Router* magazine.

Guild of Master Craftsman Publications Ltd,
Castle Place, 166 High Street, Lewes,
East Sussex BN7 1XU, United Kingdom

Tel: 01273 488005 Fax: 01273 402866
Website: www.gmcbooks.com

Contact us for a complete catalogue, or visit our website.
Orders by credit card are accepted.